Trees

Workshop in Versailles, June 14–16, 1995

Brigitte Chauvin
Serge Cohen
Alain Rouault
Editors

1996

Birkhäuser Verlag
Basel · Boston · Berlin

Editors' address:

Université de Versailles Saint-Quentin
45, Avenue des Etats Unis
78035 Versailles Cedex
France

Mathematics Subject Classification (1991): 60Jxx, 60Gxx, 82Bx, 05C05

A CIP catalogue record for this book is available from the Library of Congress,
Washington D.C., USA

Deutsche Bibliothek Cataloging-in-Publication Data

Trees : workshop in Versailles, June 14–16, 1995 / Brigitte
Chauvin ... ed. – Basel ; Boston ; Berlin : Birkhäuser, 1996
    (Progress in probability ; Vol. 40)
    ISBN 3-7643-5453-4 (Basel ...)
    ISBN 0-8176-5453-4 (Boston)
NE: Chauvin, Brigitte [Hrsg.]; GT

© 1996 Birkhäuser Verlag Basel, P.O. Box 133, CH–4010 Basel
Printed on acid-free paper produced from chlorine-free pulp. TCF ∞
ISBN 3-7643-5453-4
ISBN 0-8176-5453-4
9 8 7 6 5 4 3 2 1

Contents

When I was going to St. Ives
I met a man with seven wives
Each wife had seven cats,
Each cat had seven kits,
Each kit had seven sacks,
Sacks, kits, cats and wives,
How many were going to St. Ives?

Introduction

En Probabilités, la notion d'arbre est apparue comme une notion sous-jacente à celle des processus de branchement, processus qui ont fait l'objet de très nombreux travaux ces cinquante dernières années. Mais pendant longtemps ces arbres n'ont été étudiés que par l'intermédiaire de certaines fonctions aléatoires de comptage qui leur sont naturellement associées. Les arbres développaient l'intuition des chercheurs mais le travail technique s'effectuait à l'aide de la théorie des processus de Markov.

Le développement des travaux sur les processus markoviens de branchement, l'application des techniques approfondies des martingales à ces processus ont conduit progressivement à donner aux arbres un statut à part entière ; des arbres continus sont apparus. Dans le même temps, des études sur les fractales aléatoires et le chaos multiplicatif faisaient un usage intensif des arbres et rejoignaient en partie de cette manière les travaux précédents.

Les arbres rencontrés en informatique dans les algorithmes de tri ont donné lieu également ces dix dernières années à des travaux de nature combinatoire et probabiliste importants. Enfin, les physiciens statisticiens étudient depuis longtemps leurs systèmes désordonnés sur des structures géométriques dont les plus simples sont les arbres.

Le mérite de ce colloque a été de rassembler des spécialistes des diverses disciplines citées qui, à travers leurs contributions variées, ont pu démontrer la vitalité de leurs recherches entreprises sur les arbres.

Jacques Neveu

Editors' Preface

This volume contains the proceedings of the Workshop on Trees held in Versailles on 14–16 June 1995.

Why a workshop on trees? Two main reasons motivated this workshop. First, the current plethora of tree representations in branching processes which has become obvious by the Minneapolis IMA Congress Classical and Modern Branching Processes in 1994. But this would not have been sufficient to organize an "n-th" congress on branching processes. Secondly, regular discussions with researchers in algebra and computing sciences at the University of Versailles and at INRIA (Institut National de Recherche en Informatique et Automatique) convinced us it would be fruitful to offer the workers in these different fields the opportunity to exchange their points of view on the subject. The organizers being probabilists, a large part of the meeting (two sessions) was devoted to probability theory (not only branching processes). Nevertheless, the other three sessions focused on algorithms, on ultrametric and combinatorial aspects of trees and on disordered systems.

Most papers in this volume are both of high level and of pedagogical interest. They are intended for a large public, including graduate students looking for an initiation to tree structures. The papers have been grouped into four sections:

- disordered systems,
- probability and trees,
- large deviations,
- ultrametric and algebraic aspects of trees.

Some of the speakers are, unfortunately, not represented in this volume. They mainly contributed by their talks, remarks and questions: Philippe Biane, Bernard Derrida, Luc Devroye, Philippe Flajolet, Jean Fonlupt, Jochen Geiger, Anatole Joffe, Jean Lacroix, Jean-François Le Gall, Robin Pemantle, Yuval Peres, Bruce Reed, and Bruno Salvy,

Versailles proved to be an excellent and popular venue. An essential component was the comfortable environment offered by the new Fermat Hall.

The University of Versailles, the INRIA, the ENPC (Ecole Nationale des Ponts et Chaussées), the CNRS, the Mission Scientifique et Technique, the ONF (Office National des Forêts), by their care brought in the organisation of the meeting, largely contributed to its success.

We hope that this kind of workshop, in size midway between a highly specialized meeting and a very large congress, will be encouraged in the future.

Brigitte Chauvin
Serge Cohen
Alain Rouault

Versailles

1. Disordered Systems

Progress in Probability, Vol. 40
© 1996 Birkhäuser Verlag Basel/Switzerland

Extremality of the Disordered State for the Ising Model on General Trees

DMITRY IOFFE[*]

Mathematics Subject Classification (1991): 82B20, 82B26, 82B43

Keywords: Countable trees, Ising model, FK representation

Abstract. We develop a method to study extremality of the disordered state $\mathbb{P}^\beta$ for the Ising model on a general countable tree $\mathbf{T}$. It is shown that the tail σ- field is $\mathbb{P}^\beta$-trivial as soon as β is less than the spin glass critical inverse temperature β_c^{SG}, which is determined from the relation $\tanh(\beta_c^{SG}) = 1/\sqrt{br(\mathbf{T})}$. The method is based on the FK representation of ferromagnetic systems and recursive estimates on conditional expectations of the spin at the root. Similar estimates in the context of the bit reconstruction problem on general trees were originally obtained in [EKPS] using different methods.

Résumé. On développe une méthode pour étudier l'extrémalité de l'état désordonné $\mathbb{P}^\beta$ pour le modèle d'Ising sur un arbre dénombrable général $\mathbf{T}$. On montre que la tribu de queue est triviale dès que β est inférieur à β_c^{SG}, (inverse de la température critique du verre de spin) déterminé par la relation:

$$\tanh(\beta_c^{SG}) = 1/\sqrt{br(\mathbf{T})}.$$

La méthode est fondée sur la représentation FK des systèmes ferromagnétiques et sur des estimées récursives des espérances conditionnelles du spin à la racine. Des estimées similaires dans le contexte du problème de la reconstruction du spin sur des arbres généraux ont été obtenus d'abord dans [EKPS] par W.Evans, C.Kenyon, Y.Peres, L.Schulman en utilisant des méthodes différentes.

1 Introduction

Let $\mathbf{T}$ be a countable locally finite tree. Choose a vertex $0 \in \mathbf{T}$, which from now on will be called the root of $\mathbf{T}$. Once the root is fixed, one encounters a natural partial ordering on the set of vertices of $\mathbf{T}$. Namely, for $k, l \in \mathbf{T}$, we say that $k \prec l$ if k lies on the unique chain of edges leading from l to the root. If $k \prec l$ and k, l are nearest neighbours, then l is called a successor of k. For each $k \in \mathbf{T}$ let $b(k)$ denote the number of the successors of k, in other words, $b(k)$ is the forward branching ratio at k. It is possible [L] to introduce an average forward branching number of $\mathbf{T}$ as well: A cutset π is a subset $\pi \in \mathbf{T}$, such that any infinite selfavoiding chain of bonds emanating from the root contains exactly one vertex from π. Then,

$$br(\mathbf{T}) \overset{def}{=} \inf\{\lambda > 0 : \liminf_{\pi-cutset} \sum_{l \in \pi} \lambda^{-|l|} = 0\},$$

*)   The work was partially supported by the NSF grant DMS 9504513 and by the Commission of the European Union under the contract CHRX-CT93-0411

where $|l|$ is the number of bonds in the unique chain leading from l to the root. We refer to [L] for a comprehensive discussion of this quantity.

The purpose of this work is to develop a convenient tool for studying extremality properties of the disordered state for the Ising model on $\mathbf{T}$. This question has a somewhat curious history for homogeneous trees $\mathbf{T}_d$ with the forward branching ratio $d \in \mathbb{N}$: The critical inverse temperature β_c for the Ising model on $\mathbf{T}_d$ is given [P] by

$$\tanh(\beta_c) = 1/d .$$

In a seminal paper [S] Spitzer constructed a family of Ising Gibbs measures, which he called Markov chains on trees, and asserted that above β_c any such Markov chain is extremal, the limit state with free boundary conditions in particular. However, Higuchi in [H] reproduced an unpublished computation of Kamae, which clearly revealed that the disordered state cannot be extremal as long as

$$\tanh(\beta) > 1/\sqrt{d} .$$

An insight into the above quantity came with the work of [CCST], where they identified

$$\tanh(\beta_c^{SG}) = 1/\sqrt{d}$$

as a spin glass critical inverse temperature in the case of the binary tree $d = 2$.

Bleher [B] tried to modify their approach in order to prove that β_c^{SG} is, in fact, a threshold for extremality for the disordered state as well, but in the latter case the situation is substantially complicated due to the loss of the independence, which is intrinsic for the spin glass picture, and Bleher's proof was eventually found to be erroneous. The corresponding mistake was corrected in [BRZ] at a price of rather tedious computations. A different method, based on the FK representation, was suggested in [I]. Here we extend the techniques of [I] to derive results on general trees. These results, however, are not as satisfactory as those obtained for the Bethe lattice. Specifically, we prove

Theorem 1. *If*

$$\tanh(\beta) < 1/\sqrt{br(\mathbf{T})}, \tag{1}$$

then the disordered state at the inverse temperature β is extremal.

As in [BRZ], our strategy to prove the above theorem is based on an observation that the tail triviality of the spin at the root is equivalent to the tail triviality of the disordered state itself, which is stated in lemma 2 below, and, most crucially, on recursive estimates on second moments of conditional expectations of the spin at the root, formulated in the proposition 4 below, or, more precisely, on its weaker version (14). An estimate similar to (14) was originally proved in [EKPS] in the context of the bit reconstruction problem on general trees using completely different methods.

Now, a proper generalization [EKPS] of the computation in [H] shows that the extremality fails whenever

$$\tanh(\beta) > 1/\sqrt{\overline{br}(\mathbf{T})}.$$

What, therefore, remains unclear for us is the exact state of affairs at the critical point β_c,

$$\tanh(\beta_c) \;=\; 1/\sqrt{br(\mathbf{T})}.$$

Never the less, we hope that our estimate (10) below already provides all the necessary data to enable a complete treatment of the critical case[1].

The paper is organized as follows: In section 2 we describe the model and list some of its relevant properties. Section 3 is devoted to the FK arithmetics, which lies in the heart of our approach. Theorem 1 is proved in section 4.

2 The Model

Let $\mathbf{V}$ be a finite tree with a distinguished root 0 and the boundary

$$\partial\mathbf{V} \;=\; \{k \in \mathbf{V} : \; b(k) = 0\}.$$

Recall that $b(k)$ is the forward branching ratio at k with the respect to the root. The Ising Gibbs state on $\mathbf{V}$ at the inverse temperature β and with boundary condition ξ on $\partial\mathbf{V}$ is the probability distribution on $\Omega_V = \{-1,1\}^{\mathbf{V}}$, given by

$$\mathbb{P}^{\beta}_{\mathbf{V},\xi}(x) \;=\; \frac{1}{\mathbf{Z}^{\beta}_{\mathbf{V},\xi}}\,\exp\{\frac{\beta}{2}\sum_{|k-l|=1} x_k x_l \;+\; \beta \sum_{k \in \partial\mathbf{V}} \xi_k x_k\},$$

where the distance $|k - l|$ is defined to be the number of bonds in the chain connecting k and l and $\mathbf{Z}^{\beta}_{\mathbf{V},\xi}$ is the normalizing constant or partition function. In particular, if we pick free boundary conditions $\xi \equiv 0$, then the corresponding measure will be called the disordered state (on $\mathbf{V}$), and we shall denote it as $\mathbb{P}^{\beta}_{V}$.

For each $k \in \mathbf{V}$ let $S(k)$ to denote the set of successors of k,

$$S(k) \;=\; \{l : \; k \prec l \text{ and } |l - k| = 1\}.$$

Assume that k is such that $S(k) \subseteq \partial\mathbf{V}$. The following computation is very specific for the tree structure of $\mathbf{V}$:

$$\sum_{l \in S(k)} \sum_{x_l = \pm 1} \mathbb{P}^{\beta}_{\mathbf{V},\xi}(x) \;=\; \mathbb{P}^{\beta}_{\mathbf{V}\setminus S(k),\nu}(x), \qquad (2)$$

where $\nu_r = \xi_r$ for $r \in \partial\mathbf{V} \setminus S(k)$ and,

$$\nu_k \;=\; \frac{1}{2}\sum_{l \in S(k)} \log[\frac{e^{\beta+\xi_l} + e^{-(\beta+\xi_l)}}{e^{\beta-\xi_l} + e^{\xi_l-\beta}}].$$

---

1)   I was informed by Yuval Peres that the question of the tail triviality of the spin at the root in the critical case was completely resolved in his recent work with Robin Pemantle.

Note that the function on the right hand side above is an odd function of the boundary configuration ξ. Proceeding with the summation from the boundary to the root, one obtains an odd function $H = H(\xi_l, l \in \partial\mathbf{V})$, which describes the distribution of the spin at the root given boundary conditions ξ,

$$\mathbb{E}^{\beta}_{\mathbf{V},\xi} x_0 = \tanh(H(\xi)).$$

H can be viewed as the effective magnetic field at the root given ξ on $\partial\mathbf{V}$.

In particular, for the disordered state the effective magnetic field at the root is always zero. In fact, it is easy to see by means of the computation sketched above that for any connected subtree $\mathbf{V}' \subseteq \mathbf{V}$, the relativization of $\mathbb{P}^{\beta}_{\mathbf{V}}$ on $\Omega_{\mathbf{V}'}$ is precisely $\mathbb{P}^{\beta}_{\mathbf{V}'}$. Back to our infinite tree $\mathbf{T}$ we, therefore, readily obtain that all finite volume Gibbs states on connected subtrees of $\mathbf{T}$ with free boundary conditions are, in fact, relativizations of of a certain infinite volume measure on $\Omega_{\mathbf{T}}$, which we call the disordered state on $\mathbf{T}$ and denote as $\mathbb{P}^{\beta}$. Note that because of the relativization property one can drop the subindex "$\mathbf{V}$" in $\mathbb{P}^{\beta}_{\mathbf{V}}$ even when studying events from $\Omega_{\mathbf{V}}$ proper.

Our main objective here is to understand when the disordered state $\mathbb{P}^{\beta}$ is extremal. More precisely, for each cutset π set

$$\mathcal{F}^{\pi} = \sigma(\{x_l, k \prec l \text{ for some } k \in \pi\}).$$

Define the tail σ-field of $\Omega_{\mathbf{T}}$ as

$$\mathcal{F}_{\infty} = \bigcap_{\pi} \mathcal{F}^{\pi}.$$

Recall that $\mathbb{P}^{\beta}$ is extremal if and only if $\mathcal{F}_{\infty}$ is $\mathbb{P}^{\beta}$-trivial. Our investigation of the extremality of $\mathbb{P}^{\beta}$ is based on the following lemma, which we prove in the next section:

Lemma 2. $\mathbb{P}^{\beta}$ *is extremal iff,*

$$\lim_{\pi \to \infty} \mathbf{Var}^{\beta}[\mathbb{E}^{\beta}(x_o \,|\, \mathcal{F}_{\pi})] = 0. \tag{3}$$

By the backward martingale convergence theorem the above limit always exists. Our main effort, therefore, is to derive conditions to ensure that it is actually zero. Both the proof of the lemma and the investigation of (3) depend on a simple FK arithmetics, developed in the next section.

3   FK (Fortuin-Kasteleyn) Arithmetics

Let $(\mathbf{V}, \mathcal{E}_{\mathbf{V}})$ be a finite connected graph with the vertex set $\mathbf{V}$ and the edge set $\mathcal{E}_{\mathbf{V}}$ respectively. As before, we use $\Omega_{\mathbf{V}} = \{-1,1\}^{\mathbf{V}}$ to denote the set of spin configurations on $\mathbf{V}$. Let now $\Omega^e_{\mathbf{V}} = \{0,1\}^{\mathcal{E}_{\mathbf{V}}}$ be the set of all edge (bond) configurations on $\mathcal{E}_{\mathbf{V}}$, i.e to each bond $b =< kl >\in \mathcal{E}_{\mathbf{V}}$ we assign a number $n(b)$, which is either

1, and in this case we say that b is open, or it is zero, and then we say that b is closed. Thus, each edge configuration $n \in \Omega_{\mathbf{V}}^e$ splits $\mathbf{V}$ into disjoint union of maximal connected components, where we call two vertices k and l connected, if there is a chain of open bonds leading from k to l. Let $\mathcal{C}(n)$ to denote the number of these components. Consider now a joint probability distribution μ on $\Omega_{\mathbf{V}} \times \Omega_{\mathbf{V}}^e$, given by:

$$\mu(x, n) = \frac{1}{\mathbf{Z}(\beta)} \prod_{n(<kl>)=1} q\delta(x_k - x_l) \prod_{n(<kl>)=0} (1 - q).$$

Then [ACCN], the site marginal of μ is precisely $\mathbb{P}_{\mathbf{V}}^\beta$ with β and q related via: $\beta = -1/2 \log(1 - q)$ or $q = 1 - e^{-2\beta}$. On the other hand the edge marginal of μ is given by:

$$\mathbb{Q}_{\mathbf{V}}^\beta(n) = \frac{1}{\mathbf{Z}(\beta)} 2^{\mathcal{C}(n)} q^{\sum n(b)} (1 - q)^{\sum (1-n(b))}.$$

The disordered state $\mathbb{P}_{\mathbf{V}}^\beta$ has [ACCN], then, the following convenient representation: Choose maximal connected clusters according to $\mathbb{Q}_{\mathbf{V}}^\beta$ and paint each cluster independently into ± 1 with probability $1/2$ each.

In general $\mathbb{Q}_{\mathbf{V}}^\beta$ is a non-local dependent percolation. If, however, $\mathbf{V}$ is a tree, then things are greatly simplified [CCST]. Indeed, in the latter case the number of maximal connected components of $n \in \Omega_{\mathbf{V}}^e$ equals to one plus the number of closed bonds of n, i.e.

$$\mathcal{C}(n) = 1 + \sum_{b \in \mathcal{E}_{\mathbf{V}}} (1 - n(b)).$$

Consequently,

$$\mathbb{Q}_{\mathbf{V}}^\beta(n) = \frac{2}{\mathbf{Z}(\beta)} q^{\sum n(b)} (2(1 - q))^{\sum (1-n(b))} = p^{\sum n(b)} (1 - p)^{\sum (1-n(b))},$$

where

$$p = \frac{q}{2 - q} = \tanh(\beta).$$

Therefore, in the case of trees the random cluster measure $\mathbb{Q}_{\mathbf{V}}^\beta$ is just the usual Bernoulli bond percolation at p given above.

Let us summarize the above discussion in the form of the following algorithm of construction of $\mathbb{P}_{\mathbf{V}}^\beta$ on finite subtrees $\mathbf{V} \subset \mathbf{T}$:

Step 1. Set $p = \tanh(\beta)$ and consider an independent Bernoulli percolation on $\mathcal{E}_{\mathbf{V}}$, i.e. assign to each bond configuration $n \in \Omega_{\mathbf{V}}^e$ the probability

$$\mathbb{Q}_{\mathbf{V}}^\beta(n) = p^{\sum_{b \in \mathcal{E}_{\mathbf{V}}} n(b)} (1 - p)^{\sum_{b \in \mathcal{E}_{\mathbf{V}}} (1-n(b))}.$$

Step 2. Given a bond configuration n, paint independently each maximal connected component of $\mathbf{V}$ into $+1$ or -1 with probability $1/2$ each.

The above two-step procedure can be, using some labelling algorithm to avoid ambiguities, equally applied to construct probability measures $\mathbb{P}_{\mathbf{V}}^{\beta}$ for infinite connected subtrees $\mathbf{V} \subseteq \mathbf{T}$, in particular for $\mathbf{T}$ itself. Thus, let $\mathbb{Q}^{\beta}$ denote the independent Bernoulli percolation measure on the edges of $\mathbf{T}$ and $\mathbb{P}^{\beta}$ denote the corresponding measure on Ω_T, which is, of course, the infinite volume disordered state defined earlier. Note, by the way, that the relativization property of $\mathbb{P}^{\beta}$ becomes very transparent under the FK representation. Moreover, many quantities related to $\mathbb{P}^{\beta}$ admit a natural percolation interpretation:

Let $< \bullet >^{\beta}$ to denote the expectation under $\mathbb{P}^{\beta}$. Then,

$$< x_k x_l >^{\beta} \; = \; \mathbb{Q}^{\beta}\{k \text{ is connected to } l\}.$$

Another important example is provided by the following computation:

For any finite subset $A \subset \mathbf{T}$ set

$$\mathbf{x}_A \; = \; \prod_{j \in A} x_j.$$

Then, if $|A|$ is odd,

$$< \mathbf{x}_A >^{\beta} \; = \; 0 \; . \tag{4}$$

If $|A|$ is even, let us say that a configuration $n \in \{0,1\}^{\mathcal{E}}$ splits A evenly, if there is even number of vertices of A in each maximal connected component of n. Then,

$$< \mathbf{x}_A >^{\beta} \; = \; \mathbb{Q}^{\beta}(n : \; n \text{ splits } A \text{ evenly}) \tag{5}$$

Formulas (4) and (5) are immediate consequences of the FK representation and the relativization property of $\mathbb{P}^{\beta}$. Indeed, let $\mathcal{T}(A) = (V(A), \mathcal{E}(A))$ be the minimal connected subtree which spans A. Then,

$$< \mathbf{x}_A >^{\beta} \; = \; < \mathbf{x}_A >^{\beta}_{V(A)},$$

where $< \bullet >^{\beta}_{V(A)}$ is the expectation with respect to $\mathbb{P}^{\beta}_{V(A)}$. Simple combinatoric arguments, then, imply:

$$< \mathbf{x}_A >^{\beta} \; = \; < \mathbf{x}_A >^{\beta}_{V(A)} \; = \; 0 \; ,$$

in the odd case, and

$$< \mathbf{x}_A >^{\beta} \; = \; < \mathbf{x}_A >^{\beta}_{V(A)} \; = \; \mathbb{Q}^{\beta}_{\mathcal{E}(A)}(n : \; n \text{ splits } A \text{ evenly})$$

in the even case. Since $\mathbb{Q}^{\beta}_{\mathcal{E}(A)}$ is the relativization of $\mathbb{Q}^{\beta}$, (5) follows.

Proposition 3 (I). *Let two disjoint finite subsets $A, B \subset V$ have edge disjoint minimal spanning trees, i.e $\mathcal{E}(A) \cap \mathcal{E}(B) = \emptyset$.*
a) If both $|A|$ and $|B|$ are even, then

$$< \mathbf{x}_A \mathbf{x}_B >^\beta \; = \; < \mathbf{x}_A >^\beta < \mathbf{x}_B >^\beta . \tag{6}$$

b) If both $|A|$ and $|B|$ are odd, then for any site j, which lies on the unique chain connecting $V(A)$ to $V(B)$,

$$< \mathbf{x}_A \mathbf{x}_B >^\beta \; = \; < \mathbf{x}_A x_j >^\beta < \mathbf{x}_B x_j >^\beta . \tag{7}$$

Proof. Both formulas are consequences of (4) and (5) above and independence relations for Bernoulli percolation.

We turn now to the proof of lemma 2. Let $\mathbf{S}_n$ denote the set of sites at distance n from the root. By the Markov property it will be enough to show that $\mathbb{P}^\beta$-a.s. for any n fixed and for all $A \subseteq \mathbf{S}_n$,

$$\lim_{\pi \to \infty} \mathbb{E}^\beta(\mathbf{x}_A \,|\, \mathcal{F}^\pi) \; = \; < \mathbf{x}_A >^\beta. \tag{8}$$

We shall view conditional expectations as projections on corresponding subspaces, the scalar product being defined by $< \bullet >^\beta$, and the rules of the game being dictated by the proposition 3. Let $\mathbf{T}_k$ to be the tree growing from the site k, $\pi_k = \mathbf{T}_k \cap \pi$ and set

$$\theta(k) \; = \; \lim_{\pi \to \infty} \mathbf{Var}^\beta(x_k \,|\, \mathcal{F}^{\pi_k}).$$

Lemma 2 asserts that $\mathbb{P}^\beta$ is extremal whenever $\theta(0) = 0$. Our first observation is that

$$\theta(0) = 0 \implies \theta(k) = 0 \; \forall k \in T.$$

Indeed, by the proposition 3, for any $A \subset \mathbf{T}_k$,

$$< x_0 \mathbf{x}_A >^\beta \; = \; p^{|k|} < x_k \mathbf{x}_A >^\beta \implies \theta(0) > p^{2|k|} \theta(k).$$

Let us redefine $\mathcal{F}^\pi$ to be the σ-algebra, generated by the spins from π only. Because of the Markov property the left hand side of (3) is insensitive to such an abuse of notations.

We proceed by induction. Formula (8) is precisely $\theta(0) = 0$ for $n = 0$ and assume that (8) is true for $n = 1, 2, ..., N - 1$.

Define ϕ to be a parity function on finite sets, i.e $\phi(A) = \pm 1$ depending on whether $|A|$ is even or odd respectively. Let $\mathcal{I}_N = \{-1, 1\}^{\mathbf{S}_N}$. For each cutset π far enough from the root we split $\mathcal{F}^\pi$ into a direct sum of subspaces

$$\mathcal{F}^\pi \; = \; \bigoplus_{r \in \mathcal{I}_N} \mathcal{F}^\pi_{r,N}, \tag{9}$$

where $\mathcal{F}^{\pi}_{r,N}$ is spanned by monomials $\mathbf{x}_A$, $A \subseteq \pi$, such that $\phi(A \cap \mathbf{T}_k) = r(k)$ for all $k \in \mathbf{S}_N$. By the virtue of the proposition 3, for any $\mathbf{x}_A \in \mathcal{F}^{\pi}_{r,N}$ and $\mathbf{x}_B \in \mathcal{F}^{\pi}_{q,N}$,

$$< \mathbf{x}_A \mathbf{x}_B >^{\beta} \; = \; < \mathbf{x}_A \mathbf{x}_B \prod_{k:r(k)q(k)=-1} x_k \mathbb{E}^{\beta}(x_k \,|\mathcal{F}^{\pi_k}) >^{\beta}$$

Consequently, if $\theta(0)$, and hence all $\theta(k)$, equals to zero, then

$$\lim_{\pi \to \infty} \cos(\mathcal{F}^{\pi}_{r,N}, \mathcal{F}^{\pi}_{q,N}) \; = \; \delta(r - q).$$

Therefore, whenever $\theta(0) = 0$, one can choose a constant $c < \infty$ such that for any $u = \sum_{r \in \mathcal{I}_N} u_r \in \mathcal{F}^{\pi}$,

$$\|u\| \; \leq \; 1 \; \Longrightarrow \; \max\{\|u_r\|, \; r \in \mathcal{I}_N\} \; \leq \; c$$

independently of π. We used $\| \bullet \|$ above to denote the norm defined by the scalar product $< \bullet >^{\beta}$.

Now let $A \subseteq \mathbf{S}_N$ and define $\tilde{A}$ via

$$\tilde{A} \; = \; \{k \in \mathbf{S}_{N-1} : \; \phi(S(k) \cap A) = -1\}.$$

Note that, in a view of the proposition 3,

$$< \mathbf{x}_A >^{\beta} \; = \; p^{|A|} < \mathbf{x}_{\tilde{A}} >^{\beta}.$$

Choose now a $u \in \mathcal{F}^{\pi}$ with $\|u\| \leq 1$ and let $u = \sum u_r$ be its decomposition with respect to (9). If $r \in \mathcal{I}_N$ is such that $r(k) = -1$ for some $k \in A$, then by the proposition 3,

$$|< u_r \mathbf{x}_A >^{\beta}| \; = \; |< u_r \mathbf{x}_{A \setminus k} \mathbb{E}^{\beta}(x_k \,|\mathcal{F}^{\pi_k}) >^{\beta}| \; \leq \; c\|\mathbb{E}^{\beta}(x_k \,|\mathcal{F}^{\pi_k})\|,$$

and

$$|< u_r >^{\beta}| \; = \; |< u_r x_k \mathbb{E}^{\beta}(x_k \,|\mathcal{F}^{\pi_k}) >^{\beta}| \; \leq \; c\|\mathbb{E}^{\beta}(x_k \,|\mathcal{F}^{\pi_k})\|.$$

Otherwise, if $r(k) = 1$ for all $k \in A$,

$$< u_r \mathbf{x}_A >^{\beta} \; = \; p^{|A|} < u_r \mathbf{x}_{\tilde{A}} >^{\beta} \; = \; p^{|A|} < u_r >^{\beta} < \mathbf{x}_{\tilde{A}} >^{\beta} \; + \; o(1),$$

where the second equality follows from the induction assumption. Thus,

$$\lim_{\substack{\pi \to \infty \\ \|u\|=1}} \max \; |< u \mathbf{x}_A >^{\beta} - < u >^{\beta} < \mathbf{x}_A >^{\beta}| \; = \; 0$$

and (8) follows.

4   Proof of Theorem 1

Let $\mathbf{V} = \mathbf{V}_1 \cup \mathbf{V}_2$, where $\mathbf{V}_1$ and $\mathbf{V}_2$ are two finite edge disjoint trees growing from the common root, i.e $\mathbf{V}_1 \cap \mathbf{V}_2 = 0$. Also let $A_i = \partial \mathbf{V}_i, i = 1, 2$ and let $\mathcal{F}_1, \mathcal{F}_2$ and $\mathcal{F}$ be σ- algebras generated by the spins from A_1, A_2 and $\partial \mathbf{V} = A_1 \vee A_2$ respectively. For fixed boundary configurations $\xi_i \in \Omega_i \overset{def}{=} \{-1, 1\}^{A_i}$, $i = 1, 2$, set

$$g_i(\xi_i) \; = \; \mathbb{E}^\beta(x_0 \,|\mathcal{F}_i) \;\; and \;\; g(\xi_1, \xi_2) \; = \; \mathbb{E}^\beta(x_0 \,|\mathcal{F})$$

Proposition 4 [I] *There exists a positive constant α such that,*

$$\|g\|^2 \; \leq \; \|g_1\|^2 \; + \; \|g_2\|^2 \; - \; \alpha\|g_1\|^2\|g_2\|^2 \tag{10}$$

Moreover, one may assign α to be

$$\alpha \; = \; [\cosh(\frac{1}{2}\log\frac{2 - (1-p)^{b(0)}}{(1-p)^{b(0)}})]^{-2}, \tag{11}$$

where $b(0)$ is the branching ratio at the root in $\mathbf{V}$.

Proof. Note, first of all, that

$$g \; = \; g_1 \; + \; g_2 \; - \; g_1 g_2 g. \tag{12}$$

Indeed, let $\mathbf{x}_B \in \mathcal{F}$. Then $\mathbf{x}_B = \mathbf{x}_{B_1}\mathbf{x}_{B_2}$, where $\mathbf{x}_{B_i} \in \mathcal{F}_i$; $i = 1, 2$. There are only two symmetric possibilities to consider: either $|B_1|$ is odd and $|B_2|$ is even or the other way around. So let us assume that $|B_1|$ is odd. Then, using Proposition 3, one obtains:

$$< \mathbf{x}_B g >^\beta \; =< \mathbf{x}_{B_1} x_0 >^\beta \; < \mathbf{x}_{B_2} >^\beta \; =< \mathbf{x}_{B_1} g_1 >^\beta \; < \mathbf{x}_{B_2} >^\beta \; =< \mathbf{x}_B g_1 >^\beta$$

In the same fashion,

$$< \mathbf{x}_B g_2 >^\beta \; = \; < g_1 \mathbf{x}_{B_1} >^\beta \; < x_0 g_2 \mathbf{x}_{B_2} >^\beta \; =< g_1 g_2 g \mathbf{x}_B >^\beta$$

and (12) follows. Consequently, multiplying both sides of (12) by x_0, we obtain:

$$\|g\|^2 \; = \; \|g_1\|^2 \; + \; \|g_2\|^2 \; - \; < g^2 g_1 g_2 >^\beta.$$

Thus, it remains to show that for some $\alpha > 0$,

$$< g^2 g_1 g_2 >^\beta \; \geq \; \alpha < g_1 g_2 >^\beta.$$

According to the discussion in section 2,

$$g_i(\xi_i) \; = \; \tanh(H_1(\xi_i)) \; ; \;\; i = 1, 2;$$

and

$$g(\xi_1, \xi_2) = \tanh(H_1(\xi_i) + H_2(\xi_2)),$$

where H_1, H_2 are the effective magnetic fields at 0 given boundary conditions on A_1 and A_2 respectively.

Next, define $\Omega_{i,+}$ and $\Omega_{i,-}$; $i = 1, 2$, via

$$\Omega_{i,+(-)} = \{\xi \in \Omega_i : H_i(\xi) > (<)0\},$$

Using the $\pm$ symmetry of the model, we obtain:

$$<g^2 g_1 g_2>^\beta = 2 \sum_{g_i \in \Omega_{i,+}} (\mathbb{P}^\beta(x|_{A_1} = \xi_1, x|_{A_2} = \xi_2)\tanh^2(H_1(\xi_1) + H_2(\xi_2)) -$$

$$- \mathbb{P}^\beta(x|_{A_1} = \xi_1, x|_{A_2} = -\xi_2)\tanh^2(H_1(\xi_1) - H_2(\xi_2)))g_1(\xi_1)g_2(\xi_2).$$

Now, it is easy to see that the probabilities above can be represented as follows: For any $\xi_1 \in \Omega_1$ and $\xi_2 \in \Omega_2$,

$$\mathbb{P}^\beta(x|_{A_1} = \xi_1, x|_{A_2} = \xi_2) = \frac{1}{\mathbf{Z}}\mathbf{Z}_1(\xi_1)\mathbf{Z}_2(\xi_2)\cosh(H_1(\xi_1) + H_2(\xi_2)),$$

where $\mathbf{Z}_1$ and $\mathbf{Z}_2$ are positive even functions of boundary configurations and $\mathbf{Z}$ is the corresponding partition function. Therefore,

$$<g^2 g_1 g_2>^\beta = 2 \sum_{\xi_i \in \Omega_{i,+}} g_1(\xi_1)g_2(\xi_2)\frac{\mathbf{Z}_1\mathbf{Z}_2}{\mathbf{Z}}\mathbf{U}(\xi_1, \xi_2),$$

where

$$\mathbf{U}(\xi_1, \xi_2) = \frac{\sinh^2(H_1(\xi_1) + H_2(\xi_2))}{\cosh(H_1(\xi_1) + H_2(\xi_2))} - \frac{\sinh^2(H_1(\xi_1) - H_2(\xi_2))}{\cosh(H_1(\xi_1) - H_2(\xi_2))}.$$

After a small computation one infers that

$$\mathbf{U}(\xi_1, \xi_2) \geq \frac{\cosh(H_1(\xi_1) + H_2(\xi_2)) - \cosh(H_1(\xi_1) - H_2(\xi_2))}{\cosh(H_1(\xi_1) + H_2(\xi_2))\cosh(H_1(\xi_1) - H_2(\xi_2))}. \tag{13}$$

However, due to the monotonicity of the expectation of the spin at the root in boundary conditions, one may estimate:

$$\tanh(H_1(\xi_1) + H_2(\xi_2)) \leq 1 - (1 - p)^{b(0)},$$

where the right hand side above is the expectation at the root conditioned on the event that all the spins in $S(0)$ are up. Consequently, the inverse of the denominator in (13) can be bounded above by α, specified in (11). Thus,

$$<g^2 g_1 g_2>^\beta \geq 2\alpha \sum_{\xi_i \in \Omega_{i,+}} g_1 g_2 \frac{\mathbf{Z}_1\mathbf{Z}_2}{\mathbf{Z}}(\cosh(H_1 + H_2) - \cosh(H_1 - H_2))$$

$$= \alpha <g_1 g_2>^\beta = \alpha\|g_1\|^2\|g_2\|^2,$$

and the proof of the Proposition 4 is concluded.

It is very easy now to prove the extremality of $\mathbb{P}^\beta$ in the case $\tanh(\beta) < 1/\sqrt{br(\mathbf{T})}$. Indeed, for each $k \in \mathbf{T}$ one can view $\mathbf{T}_k$ as the union of $b(k)$ disjoint trees with a common root k. Therefore, using (10) $(b(k) - 1)-$ times and taking into account that for any $l \in S(k)$ and any cutset π,

$$\mathbb{E}^\beta(x_k \,|\mathcal{F}^{\pi_l}) \;=\; p\mathbb{E}^\beta(x_l \,|\mathcal{F}^{\pi_l}),$$

we obtain that for all $k \in \mathbf{T}$,

$$\theta(k) \;\leq\; p^2 \sum_{l \in S(k)} \theta(l),$$

where $\theta(\bullet)$ was defined in section 3. Consequently[2], for each cutset π,

$$\theta(0) \;\leq\; \sum_{k \in \pi} p^{2|k|}\theta(k) \;\leq\; \sum_{k \in \pi} p^{2|k|}. \tag{14}$$

Taking a lim inf as $\pi \to \infty$, we obtain that $\theta(0) = 0$, which , by lemma 2, implies extremality.

Remark: Note that we did not use (10) in its full strength to prove the theorem for noncritical β. As in the case of Bethe lattices the negative term in the right hand side of (10) should become crucial, though, for a treatment of the critical case $\tanh(\beta) = 1/\sqrt{br(\mathbf{T})}$. For example, if $\mathbf{T}$ is spherically symmetric, i.e if $b(k)$ depends only to the distance from k to the root; $b(k) = b(|k|)$, then also $\theta(k) = \theta(|k|)$, and (10) is easily seen to yield:

$$\theta(n) \;\leq\; b(n)p^2\theta(n+1)(1 - p^2\frac{b(n) - 1}{b(n)}\theta(n+1)), \;\; n = 1, 2, \ldots .$$

Consequently, setting $M_n = p^{2(n+1)} \prod_{k \leq n} b(k)$, we obtain:

$$\theta(0) \;\leq\; M_n\theta(n+1) \prod_0^n (1 \,-\, p^2\frac{b(k) - 1}{b(k)}\theta(k+1)).$$

Therefore, $\mathbb{P}^\beta$ is clearly extremal as soon as

$$\limsup_{n \to \infty} M_n \;<\; \infty, \tag{15}$$

However, the (15) above should not be the right criteria for extremality in the spherically symmetric case. The right criteria is likely to be provided by the computation in the spirit of [H], which reveals that if

$$\sum_{n=1}^\infty \frac{1}{M_n} \;<\; \infty,$$

then the disordered state is not extremal. We failed to close this gap while preparing this report.

---

2)   see [EKPS] for a similar inequality, derived via different methods

References

[ACCN] M. Aizenman, J. T. Chayes, L. Chayes, C. M. Newman (1988), *Discontinuity of the magnetization in one-dimensional $1/|x-y|^2$ Ising and Potts models,* J. Stat. Phys. 50, 1, 1–40.

[B] P. M. Bleher (1990), *Extremity of the disordered phase in the Ising model on the Bethe lattice,* Comm. Math. Phys. 128, 411–419.

[BRZ] P. M. Bleher, J. Ruiz, V. A. Zagrebnov (1995), *On the purity of the limiting Gibbs state for the Ising model on the Bethe lattice,* J. Stat. Phys. 79, 1/2, 473–482.

[CCST] J. T. Chayes, L. Chayes, J. P. Sethna, D. J. Thouless (1986), *A mean field spin glass with short range interactions,* Comm. Math. Phys. 106, 41–89.

[EKPS] W. Evans, C. Kenyon, Y. Peres, L. Schulman (1995), *Broadcasting on trees and the Ising model,* preprint.

[H] Y. Higuchi (1977), *Remarks on the limiting Gibbs states on a (d+1)-tree,* Publ. RIMS Kyoto, 13, 335–348.

[I] D. Ioffe (1995), *A note on the extremality of the disordered state for the Ising model on the Bethe lattice,* to appear in Let. Math. Phys.

[L] R. Lyons (1990), *Random walks and percolation on trees,* Ann. Prob., 18, 3, 931–958.

[P] C. J. Preston (1974), *Gibbs States on Countable Sets,* Cambr. Univ. Press.

[S] F. Spitzer (1975), *Markov random fields on an infinite tree,* Ann. Prob., 3, 3, 387–398.

Dmitry Ioffe
WIAS
Mohrenstr. 39
D-10117 Berlin

Progress in Probability, Vol. 40
© 1996 Birkhäuser Verlag Basel/Switzerland

Trees in the Time-Scale Domain

STÉPHANE JAFFARD

Mathematics Subject Classification: 26A15, 26A30, 42A60

Keywords: Wavelets bases, trees, multifractal functions, chirps.

Abstract. We show that multifractal functions are often characterized by large wavelet coefficients on trees in the time-scale half-plane. Conversely, one can construct easily model cases of multifractal functions by imposing large wavelet coefficients on such trees.

Resumé. Nous étudions certaines séries lacunaires d'ondelettes qui ont leur coefficients placés sur des arbres du demi-espace temps-échelle. Nous montrons que ce sont des exemples de fonctions multifractales composées de singularités oscillantes.

1   Introduction

Let F be a function defined on $\mathbb{R}$. One associates with F a two-variables function, its wavelet transform defined in the "time-scale" open half-plane by

$$C(a,b) = \frac{1}{a} \int F(t)\psi(\frac{t-b}{a})dt \quad a > 0, b \in \mathbb{R} \tag{1}$$

(here ψ is smooth, compactly supported, and has a vanishing integral).

It may seem inefficient to represent a one-variable function F with the help of a two-variables function $C(a,b)$; but this may be useful for two reasons:

- Regularity properties of a function F can be translated into simple size estimates of its wavelet transform.
- If the construction of F follows a certain rule, it often happens that the construction rule is hidden on a direct inspection of F but revealed on its wavelet transform (examples such as invariant measures of dynamical systems or D.L.A. can be found in [1]) The analysis of selfsimilarfunctions in Section 2 is another example. This is similar to theuse of phase-space for dynamical systems: the introduction of an additional dimension reveals the dynamic.

Let us give examples of properties of a function F that can be translated into simple size estimates of its wavelet transform. If ψ is C^K and has K first vanishing moments, F belongs to the Sobolev space H^s for $s < K$ if and only if

$$\int\int |C(a,b)|^2 (1+|a|^2)^s \frac{dadb}{a^2} < \infty.$$

An even simpler example is supplied by uniform Hölder regularity; $F \in C^s(\mathbb{R})$ if and only if

$$|C(a,b)| \leq Ca^s.$$

If a is small, (1) takes only into account the values of F close to b; thus it is not surprising that local properties of a function can be studied on the wavelet transform. This is for instance the case for pointwise regularity.

Suppose that h is a positive real number; a function $F : \mathbb{R} \to \mathbb{R}$ is $C^h(x_0)$ if there exists a polynomial P of degree less than h such that

$$|F(x) - P(x - x_0)| \leq C|x - x_0|^h. \tag{2}$$

The following proposition of [4] relates pointwise regularity with decay conditions of the wavelet coefficients.

Proposition 1. *Let $F : \mathbb{R} \to \mathbb{R}$ be a bounded function. If F is $C^h(x_0)$,*

$$|C(a,b)| \leq Ca^h \left(1 + \frac{|b - x_0|}{a}\right)^h. \tag{3}$$

Conversely suppose that there exists $\varepsilon > 0$ such that $F \in C^\varepsilon(\mathbb{R}^d)$. If (3) holds, there exists a polynomial P of degree at most $[h]$ such that, if $|x - x_0| \leq 1/2$,

$$|F(x) - P(x - x_0)| \leq C|x - x_0|^h |\log(|x - x_0|)|. \tag{4}$$

In the following, the point x_0 is fixed, and we suppose that there exists $\varepsilon > 0$ such that $F \in C^\varepsilon(\mathbb{R}^d)$. The *Hölder exponent* of F at x_0 is the supremum of all h such that F is $C^h(x_0)$.

We introduce some notations that fit this notion. The first one is a weak form of the $\mathcal{O}$ notation of Landau, and the second one expresses the fact that two functions are of the same order of magnitude, disregarding "low order corrections".

Definition 1. *If F and G are two functions, $F = \overline{\mathcal{O}}(G)$ if*

$$\limsup \frac{\log |F|}{\log |G|} \leq 1,$$

and $F \sim G$ if

$$\lim \frac{\log |F|}{\log |G|} = 1.$$

Let h be the Hölder exponent of F at x_0; the following corollary is a direct consequence of the definition of the Hölder exponent and of Proposition 1.

Corollary 1. *Suppose that $F \in C^\varepsilon(\mathbb{R}^d)$ for an $\varepsilon > 0$. The Hölder exponent of F at x_0 is h if and only if the two following conditions hold:*
* *In the neighborhood of $(a,b) = (0, x_0)$*

$$|C(a,b)| = \overline{\mathcal{O}}(a^h + |b - x_0|^h). \tag{5}$$

- *There exists a sequence $(a_n, b_n) \to (0, x_0)$ satisfying*

$$|C(a_n, b_n)| \sim a_n^h + |b_n - x_0|^h. \tag{6}$$

We will call such a sequence (a_n, b_n) a *minimizing sequence* for F at x_0. If F has only one Hölder singularity at a point x_0 and is smooth elsewhere, we will thus find (at least) one sequence (a "branch") of large wavelet coefficients $(a_n, b_n) \to (0, x_0)$. Suppose now that the Hölder exponent of F changes from point to point. This situation will often correspond to a whole tree of large wavelet coefficients in the time-scale half-plane, and the branches of this tree will tend to each point of the singular support of F.

At this point, let us define what is a tree in the time-scale half-plane. The tree-top is an arbitrary point $(a_\emptyset, b_\emptyset)$. The whole tree is indexed by all n-tuples $(i_1, \ldots, i_n) \in \{1, \ldots, s\}^n$, $n \in \mathbb{N}$, and satisfies, if $i = (i_1, \ldots, i_n)$ and $i' = (i_1, \ldots, i_n, i_{n+1})$,

$$\left. \begin{array}{l} a_{i'} < a_i \le C a_{i'} \quad \text{for} \quad C > 1 \\[2mm] |b_i - b_{i'}| \le C a_i. \end{array} \right\} \tag{7}$$

Condition (7) means that the distance between a branch $i = (i_1, \ldots, i_n)$ and its children $(i_1, \ldots, i_n, i_{n+1})$ is of the order of magnitude of a constant, using the hyperbolic distance in the half-plane. (Note that, for this distance the points of the line $a = 0$ are at an infinite distance from a given point of the half-plane).

Standard examples of functions whose Hölder exponent changes from point to point are supplied by *multifractal functions*. In order to give an idea of what multifractal functions are, we first define the *Hölder spectrum* of F, which is the function $d(h)$ defined for each $h \ge 0$ as follows:
$d(h)$ is the Hausdorff dimension of the set of points x_0 where the Hölder exponent of F is h.

A multifractal function is a function which has a "nontrivial" Hölder spectrum, i.e., is a function for which $d(h)$ is defined and nonconstant on an interval $[h_{min}, h_{max}]$. Multifractal functions actually have large wavelet coefficients on a tree in the time-scale half-plane.

2   Selfsimilar functions

Examples of multifractal function are supplied by the *selfsimilar functions* (see [4]) which we now define on $\mathbb{R}$.

Definition 2. *A function F is selfsimilar if the three following conditions hold:*
- *There exists an open interval Ω and contractive similitudes $S_1, \ldots, S_s$ such that*

$$S_i(\Omega) \subset \Omega \tag{8}$$

$$S_i(\Omega) \cap S_j(\Omega) = \emptyset \quad \text{if} \quad i \ne j. \tag{9}$$

(The S_i are the product of a translation with the mapping $x \to \nu_i x$ where $\nu_i < 1$.)

- *There exists a C^∞ function g such that g and its derivatives of order less than k have fast decay and F satisfies*

$$F(x) = \sum_{i=1}^{s} \mu_i F(S_i^{-1}(x)) + g(x). \tag{10}$$

- *The function F is not uniformly C^∞ in a certain closed subset of Ω.*

Let us start with some preliminary results concerning the mappings S_i. If A is a subset of $\mathbb{R}$, let us define the mapping S by

$$S(A) = \bigcup_{i=1}^{s} S_i(A);$$

and let K be the set defined by

$$K = \bigcap_{n \in \mathbb{N}} S^n(\bar{\Omega}).$$

K is called the invariant compact set of S; its Hausdorff dimension is the number d_{max} satisfying

$$\sum_i \nu_i^{d_{max}} = 1.$$

Let i be a finite sequence $i = (i_1, \ldots, i_n)$. We define $x_i = S_{i_1} \ldots S_{i_n}(0)$, and if the sequence i is infinite, $x_i = \lim_{n \to \infty} x_{(i_1, \ldots, i_n)}$. Similarly, let $\nu_i = \nu_{i_1} \ldots \nu_{i_n}$ and $\mu_i = \mu_{i_1} \ldots \mu_{i_n}$. Thus, to each sequence $i \in \{1, \ldots, s\}^{\mathbb{N}}$ we associate a unique point x_i in K. This correspondence is in general not one to one (consider for instance the example of the lacunary trigonometric series $\sum_{n=0}^{\infty} \mu^n \sin(2^n x)$ whichare clearly selfsimilar and for which the dyadic points are limit of two sequences).

The points of K can also be represented as the limit points of the branches of the following tree T constructed in the time-scale half-plane. The tree-top is conventionally the point $(0, 1) \in \mathbb{R} \times \mathbb{R}^+$. This tree-top is linked to the s first nodes, which are the $(S_j(0), \nu_j)$. This point $(S_j(0), \nu_j)$ is linked to the $(S_j S_k(0), \nu_j \nu_k), \ldots$. If $\mathbb{R}$ is identified to $\mathbb{R} \times \{0\}$, clearly, the branch indexed by a sequence $i \in \{1, \ldots, s\}^{\mathbb{N}}$goes towards the point $(x_i, 0)$ (and it is the only one which does so if the mapping $i \to x_i$ is one-to-one). This tree is precisely related to the wavelet transform of F: in [4] we prove that there exists a small rectangle $[x_i - C\nu_i, x_i + C\nu_i] \times [C'\nu_i, C''\nu_i]$ centered at (x_i, ν_i) ($C, C'' > 1$, $C' < 1$) such that the supremum on the wavelet transform of F on this rectangle is equivalent to $|\mu_i|$ (when the length of i tends to ∞).Thus, the wavelet transform is schematically a deterministic multiplicative process along this tree :When one goes from the point indexed by $(i_1, \ldots, i_n)$ to the point indexed by $(i_1, \ldots, i_n, i_{n+1})$, the wavelet transform is multiplied by $\mu_{i_{n+1}}$. Using Corollary 1, one can then compute the Hölder exponent of these functions at every point and deduce that they are multifractal, their spectrum being obtained as follows. Let

$$h_{min} = \inf_{i=1,\ldots,s} \left(\frac{\log \mu_i}{\log \nu_i} \right), \quad h_{max} = \sup_{i=1,\ldots,s} \left(\frac{\log \mu_i}{\log \nu_i} \right);$$

(we use this notation because h_{min} will turn out to be the smallest pointwise Hölder regularity exponent of F and h_{max} the largest). Let τ be the function defined by

$$\sum_{i=1}^{s} \mu_i^{a} \nu_i^{-\tau(a)} = 1.$$

Proposition 2. *The spectrum $d(h)$ vanishes outside $[h_{min}, h_{max}]$, and on this interval,*

$$d(h) = (\inf_{a} ah - \tau(a)) \tag{11}$$

These spectra have roughly the shape of the upper half of an ellipse. This is actually the generic shape of the Hölder spectra of most multifractal signals (though we do not have reasons to believe that they follow such a simple selfsimilarity condition).

3   Oscillating singularities

Selfsimilar functions have large wavelet coefficients disposed on a tree in the time-scale half-plane. Conversely, one can construct multifractal functions by directly imposing the size of their wavelet transform at some points of the time-scale half-plane disposed on a tree. Since it is difficult to impose the size of a wavelet transform at a given point, it is easier to use an *orthonormal wavelet basis*

$$\psi_{j,k}(x) = \psi(2^j x - k), \; j, k \in \mathbf{Z}$$

where the wavelet ψ is compactly supported, C^K, has a vanishing integral, and K first vanishing moments, for K "large enough" (see [3]). We denote by $B \, (= B_{j,k})$ the dyadic interval $[k2^{-j}, (k+1)2^{-j})$ where the wavelet is centered. We will use these intervals to index the wavelets, so that ψ_B is the wavelet "centered" on the interval B. Note that we use an L^∞ normalization for the wavelets, so that the wavelet coefficients are given by

$$C_B = 2^j \int F(x)\psi_B(x)dx.$$

The use of a wavelet basis allows to impose exactly the value of the wavelet transform at the points $(a, b) = (2^{-j}, k2^{-j})$ $(j, k \in \mathbf{Z})$. The technique of fixing large wavelet coefficients on a tree allows to construct functions with very specific local behavior, and in particular functions which exhibit a very oscillatory "chirp-like" behavior. Let us start with a loose definition.

A chirp of type (h, β) should oscillate

$$f(x) = |x - x_0|^h \sin(\frac{1}{|x - x_0|^\beta}) \tag{12}$$

in the neighborhood of x_0. In signal analysis, this notion should cover functions whose *instantaneous frequency* increases fast at some time.

A remarkable property of the function (12) is that after n integrations, the function obtained has Hölder regularity $h + n(\beta + 1)$ at x_0 (the gain in regularity is not 1 at each integration, as might be expected, but $\beta + 1$). We can use this property in order to define an *oscillating singularity* at x_0 (see also [6] for another definition).

Let $h^t(x_0)$ denote the Hölder exponent of the fractional integral of order t at x_0 of a function F. More precisely, if F is a bounded function, we denote by $h^t(x_0)$ the Hölder exponent at x_0 of a function such that

$$F^t = (-\Delta)^{-t/2}(\phi F) \tag{13}$$

ϕ is a C^∞ compactly supported function satisfying $\phi(x_0) = 1$. If one performs a fractional integration of order t of function (12) then $h^t(x_0) = h + (1 + \beta)t$. We see that the gain of pointwise Hölder regularity at x_0 after a fractional integration of small order t is $(1 + \beta)t$; hence the following definition for *oscillating singularity exponents*.

Definition 3. *Let $F : \mathbb{R}^d \to \mathbb{R}$ be a bounded function. The oscillating singularity exponents of F at a point x_0 are defined by*

$$(h, \beta) = \left(h(x_0), \left. \frac{\partial}{\partial t} h^t(x_0) \right|_{t=0} - 1 \right) \tag{14}$$

These exponents belong to $[0, +\infty] \times [0, +\infty]$.

Remark: This definition makes sense because, if F is bounded x_0 is fixed, the function $t \to h^t(x_0)$ is right-differentiable (with a possible derivative of $+\infty$), see [2].

The following result shows how to compute the oscillating singularity exponents from the wavelet transform.

Proposition 3. *Let $F \in C^\varepsilon(\mathbb{R}^d)$ for an $\varepsilon > 0$. The oscillating singularity exponents at x_0 are (h, β)*

- $|C(a, b)| = \overline{\mathcal{O}}(a^h + |b - x_0|^h)$ *in the neighborhood of $(a, b) = (0, x_0)$,*
- *there exists a sequence $(a_n, b_n) \to (0, x_0)$ such that*

$$a_n + |b_n - x_0| \sim a^{1+\beta} \quad \text{and} \quad |C(a_n, b_n)| \sim a_n^h + |b_n - x_0|^h \tag{15}$$

- β *is the smallest number such that (15) holds.*

This result is a complement of Corollary 1. It relates the local geometry of the branch of large coefficients with the oscillation exponent β.

We denote by $E_{h,\beta}$ the set of points where the oscillating singularity exponents are (h,β), and by $d(h,\beta)$ the Hausdorff dimension of $E_{h,\beta}$.

Let us now construct functions with oscillating singularities on a Cantor set. This construction is a joint work with A. Arneodo, E. Bacry and J. F. Muzy [2]. These functions are constructed by assigning nonvanishing orthonormal wavelet coefficients on a tree in the time-scale half-space. But in this case, the branches will be almost parallel to the b-axis; this particular feature (different from the selfsimilar case) will creat eoscillating singularities.

Consider the usual "tree" construction of a Cantor set:

- At level 0, we start with the interval $A_\emptyset = [0,1]$.
- At level 1, we construct s disjoint intervals $A_1, \ldots, A_s$ included in $[0,1]$, where $A_k = [x_k, y_k]$ has length $|A_k| = y_k - x_k = \nu_k$.
- At level n, we have s^n subintervals A_i indexed as usual by an n- tuple $i = (i_1, \ldots, i_n) \in \{1, \ldots, s\}^n$; $A_i = [x_i, y_i]$ and the length of A_i is

$$|A_i| = \nu_i = \nu_{i_1} \ldots \nu_{i_n}.$$

We denote by K the Cantor set

$$K = \bigcap_{n \in \mathbb{N}} \left(\bigcup_{i = \in \{1, \ldots n\}^n} A_i \right).$$

We pick s ratios for the scales $\lambda_1, \ldots, \lambda_s$ of the intervals indexing the wavelets such that

$$\forall k = 1, \ldots s, \quad \lambda_k \leq \nu_k;$$

then, we construct on each interval A_i a dyadic subinterval B_i of length

$$|B_i| = \lambda_i = C(B_i)\lambda_{i_1} \ldots \lambda_{i_n}$$

where $C(B_i) \in [1,2)$ is chosen such that λ_i is a power of 2.

We still have to define the exact position of the wavelets and the size of the wavelet coefficients. For that we pick $\eta \in [0,1]$ and the interval B_i is the unique dyadic interval of length λ_i satisfying the condition

$$x_i + \eta \nu_i \in B_i$$

(it is thus a subinterval of A_i). We make in the following the asumption:

$$\text{Either } \eta \notin K \text{ or } (\lambda_1, \ldots, \lambda_s) = (\nu_1, \ldots, \nu_s).$$

In order to define the wavelet coefficients, we pick s ratios $\mu_1, \ldots, \mu_s$ in $[0,1)$, and the corresponding wavelet coefficient of the wavelet indexed by B_i is

$$\mu_i = \mu_{i_1} \ldots \mu_{i_n}.$$

The function F is completely determined by setting to zero all other wavelet coefficients.

In order to give the pointwise oscillation exponents at each point, we have to give some definitions.

Definition 4. *Let (a_n) be a sequence of real numbers. A minimizing subsequence for (a_n) is a subsequence n_k such that*

$$\liminf_{n \to \infty} a_n = \lim_{n \to \infty} a_{n_k} \tag{16}$$

We denote by $S(a_n)$ the set of all minimizing subsequences for (a_n); and if (a_n) and (b_n) are sequences, we define

$$\liminf_{(a_n)} b_n = \inf_{\{(n_k) \in S(a_n)\}} \liminf b_{n_k} \tag{17}$$

where the infimum is taken over all minimizing subsequences (n_k) for (a_n).

For instance if $a_n = (-1)^n$, $\displaystyle\liminf_{(a_n)} b_n = \liminf b_{2n+1}$.

As in the case of selfsimilar functions, the set of indexes i is interpreted as a tree and a *branch above* x_0 is a sequence i such that

$$dist(x_0, A_i) \le C\nu_i \tag{18}$$

where the constant C will be fixed later.

Proposition 4. *Let $x_0 \in K$; F has at x_0 the oscillating singularity exponents*

$$(h, \beta) = \left(\liminf \left[\frac{\log \mu_i}{\log \nu_i} \right], \liminf_{(\log \mu_i / \log \nu_i)} \left[\frac{\log \lambda_i}{\log \nu_i} \right] - 1 \right) \tag{19}$$

where the liminf is taken over all branches above x_0.

We define the function τ by

$$\sum_{l=1}^{s} \lambda_l^a \mu_l^b \nu_l^{-\tau(a,b)} = 1.$$

The following theorem gives the *Spectrum of oscillating singularities* of F.

Theorem 1. *The Hausdorff dimension $d(h, \beta)$ of the set $E_{h,\beta}$ is given by*

$$d(h, \beta) = \inf_{a,b}(a(\beta + 1) + bh - \tau(a,b)). \tag{20}$$

4 Lacunary wavelet series

The functions considered in the previous section have very few nonvanishing wavelet coefficients; they are thus examples of *lacunary wavelet series*. One might believe that the oscillating singularities are created because of the very specific position of the nonvanishing wavelet coefficients along trees in the time-scale half-plane. This is in fact not the case, as the following random example will show (where no correlations will exist between coefficients at different scales).

The function F $(= F^{\alpha,\beta})$ is defined as follows. Let $\eta < 1$. For each $j \geq 0$ we choose at random and independently $[2^{\eta j}]$ wavelet coefficients which take the value $[2^{\alpha j}]$, and we set to 0 all other wavelet coefficients. This is the most elementary model of random lacunary wavelet series one can think of.

Theorem 2. *The function F whose wavelet coefficients are defined as above has almost surely almost everywhere an oscillating singularity of exponents*

$$(h, \beta) = (\frac{\alpha}{\eta}, \frac{1}{\eta} - 1).$$

Proof. For each j denote by E_j the set of k's such that the wavelet coefficient $C_{j,k}$ is not vanishing. Let $\delta \in [0, 1]$. Denote by $I_{j,k}$ the interval centered at $k2^{-j}$ $(k \in E_j)$ and of length $2^{-\delta j}$.

First Case: $\delta > \eta$. In that case

$$\mathbb{P}(x \in \bigcup_k I_{j,k}) \leq \sum_k \mathbb{P}(x \in I_{j,k}) \leq 2^{(\eta-\delta)j}$$

so that, using Borel-Cantelli lemma,

$$\mathbb{P}(x \in \limsup_{j \to \infty} \bigcup_k I_{j,k}) = 0.$$

Second Case: $\delta \leq \eta$. In that case

$$\mathbb{P}(x \notin I_{j,k}) = 1 - 2^{-\delta j}$$

so that

$$\mathbb{P}(x \notin \bigcup_{k \in E_j} I_{j,k}) = (1 - 2^{-\delta j})^{[2^{\eta j}]}.$$

Thus

$$\mathbb{P}(x \in \bigcup_{k \in E_j} I_{j,k}) \geq 1 - C2^{(\eta-\delta)j}$$

so that using the Borel-Cantelli lemma, almost every x belongs to $\limsup_{j \to \infty} \bigcup_k I_{j,k}$,

hence we have proved the theorem.

Remark: These lacunary wavelet series turn out to be another very simple example of multifractal functions composed of chirps. We will perform a complete multifractal analysis of this function in a forthcoming paper [5].

References

[1] A. Arneodo, E. Bacry J. F. Muzy *The thermodynamics of fractals revisited with wavelets* Physica A Vol. 213 p. 232–275 (1995).

[2] A. Arneodo, E. Bacry, S. Jaffard and J. F. Muzy *Oscillating singularities on Cantor sets part I and II* (manuscript).

[3] I. Daubechies. *Orthonormal bases of compactly supported wavelets.* Comm. on pure and Appl. Math. Vol. 41(1988).

[4] S. Jaffard *Multifractal Formalism for functions Part I and II.* To appear in SIAM Journal on Mathematical Analysis.

[5] S. Jaffard *On lacunary wavelet series.* In preparation.

[6] S. Jaffard and Y. Meyer *Wavelet methods for pointwise regularity and local oscillations of functions* To appear in Memoirs of the A.M.S.

Stéphane Jaffard
Département de Mathématiques
Université Paris XII, Créteil

and

C.M.L.A., E.N.S. Cachan
61 av. du Président Wilson
94235 Cachan Cedex
France

Progress in Probability, Vol. 40

Random Measures on Trees and Thermodynamic Formalism

F. KOUKIOU*

Mathematics Subject Classification: 60G57, 60G30, 60G42

Keywords: random measures, multiplicative chaos, phase transitions

Abstract. Progress made recently on the understanding of phase transitions of disordered systems is reviewed (joint work with P. Collet). Such understanding relies on the use of the thermodynamic formalism for the study of singularities of some random measures defined by iterated multiplications.

Résumé. L'étude, à l'aide du formalisme thermodynamique, de singularités locales de certaines mesures aléatoires et la relation avec le comportement critique de systèmes désordonnés est présentée brièvement (travail en collaboration avec P.Collet).

1　Modelling the randomness

One of the paradigms of modern mathematical physics is randomness. In order to model it, one has to add to the Hamiltonian either a random interaction between the components of the system, or an external random field. These random parameters, corresponding to the static nature of the disorder in the physical system, are often called quenched. In the case of an external random field, a natural question one can address concerns the stability of phase transitions in ordered systems under random perturbations. The study of the case of random interactions between the components of the system is much more difficult. For most of the physically interesting models and in spite of many efforts, conjectures, and numerical simulations there is no rigorous proof about the existence of some kind of phase transition. In this situation, mean field models defined on homogeneous graphs provide useful tractable models for the study of many qualitative aspects of phase transitions. We shall use in the following the thermodynamic formalism [9] to study a class of random measures which is naturally connected to the mean field models of spin glasses and directed polymers indroduced by physicists.

2　Random measures on trees

The random measures we shall consider, defined by iterated multiplications, are introduced first by Mandelbrot [7] in a discontinuous model for turbulence, and studied in detail by Kahane and Peyrière in terms of Multiplicative Chaos [5], Guivarc'h [4], and more recently by Waymire and Williams [8]. Another, closely related, important class of measures, namely the branching random walk, is studied by Biggins and Kyprianou [1] and Chauvin and Rouault [2].

---

*)　Work partially supported by the EU grant CHRX-CT93-0411

Let c be an integer ≥ 2 and W a real bounded random variable defined on $(\Omega, \mathcal{F}, P)$ of mean $E(W) = 1$. Consider the finite rooted c-tree $\mathcal{T}_n$ (*i.e.* each vertex has c edges). It is an easy matter to check that $\mathcal{T}_n$ is one-to-one correspondance with the n-th generation of the c-adic partition of the unit interval $[0, 1]$:

$$\mathcal{T}_n = \bigcup_{p=1}^{n} \{[kd^{-p}, (k+1)d^{-p}[, \ \ 0 \leq k \leq d^p - 1\}.$$

We can now define random measures on $\mathcal{T}_n$ by the following way. To each atom I of the partition $\mathcal{D}_n$, we associate the random variable W_I, having common distribution with W. We further assume that for $J \cap I = \emptyset$, the corresponding variables W_I and W_J are independent. We have thus a sequence of random variables $\{W_J, J \in \mathcal{D}_p\}$, indexed by the edges $\mathcal{T}_n$. A random measure on $\mathcal{T}_n$ can be defined as

$$\mu_n^\beta(I) = c^{-n}(E(W^\beta))^{-n} \prod_{p=1}^{n} \sum_{J \in \mathcal{D}_p} W_J^\beta \lambda(J \cap I) c^n,$$

where λ denotes the Lebesgue measure. The parameter $\beta \in R$ is called inverse temperature. Such a random measure, measurable with respect to the σ-field generated by the family $\{W_J, J \in \mathcal{D}_p\}$, defines a positive martingale. Moreover it associates a random weight to each of the c^n "paths" of $\mathcal{T}_n$. This weight can be interpreted as a Gibbs factor if one considers each path as a spin configuration or as the energy of directed random walk over $\mathcal{T}_n$.

For $\beta > 0$, the "finite volume" partition function $Z_n(\beta)$ is given by the positive martingale

$$Z_n(\beta) = \sum_{I \in \mathcal{D}_n} \mu_n^\beta(I).$$

Another important thermodynamic quantity is the pressure given by

$$P_n(\beta) = n^{-1} \log Z_n(\beta).$$

One can also define the Gibbs "ensemble" for $\mathcal{T}_n$ i.e. the random probability measure

$$\nu_n^\beta(\cdot) = Z_n^{-1}(\beta)\mu_n^\beta(\cdot)$$

The name of *thermodynamic limit* (the macroscopic limit) is given to a limit when $n \to \infty$.

This general scheme is encountered in the definitions of the mean field models studied by physicists.

In the following section we shall study the behaviour of the previous thermodynamic quantities at the thermodynamic limit as a function of $\beta > 0$. The case of $\beta = 1$ is given in [5, 4, 8]. In a recent work we generalise the formalism for complex β [6].

3 The critical behaviour

In this section we briefly present the results on the phase transition of the models defined previously. The detailed proofs can be found in [3].

Let β_c be the critical temperature defined by

$$\frac{\beta_c}{E(W^{\beta_c})} E(W^{\beta_c} \log_c W) = 1 + \log_c E(W^{\beta_c}).$$

In the high temperature phase, defined by $\beta < \beta_c$, we have the following results for the thermodynamic quantities.

Theorem 3.1 *For* $0 < \beta < \beta_c$, *the limit*

$$P(\beta) = \lim_{n \to \infty} n^{-1} \log_c Z_n(\beta)$$

exists almost surely, it is nonrandom and it is given by $\log_c E(W^\beta) - \beta + 1$.

In this situation we say that the pressure (or the free energy) is a selfaveraging quantity. This property, in physics, gurantees the reproductibility of the measurements performed on a large number of physical samples. For the Gibbs ensemble we have the

Theorem 3.2 *For* $0 < \beta < \beta_c$, *the probability measures* $\nu_n^\beta(\cdot)$ *converge almost surely as* $n \to \infty$, *to a unique limit* $\nu^\beta(\cdot)$ *on the Borel field of* $[0,1]$.

This result provides a definition of the high temperature state for a mean field random model. We can also estimate the Hausdorff dimension d_H of the probability measure ν_β;

$$d_H = 1 + \log_c E(W^\beta) - \frac{\beta}{E(W^\beta)} E(W^\beta \log_c W).$$

The study of the low temperature phase, defined by $\beta \geq \beta_c$, is more difficult. One first result concerns the existence of the thermodynamic limit of the pressure and it is given by the

Theorem 3.3 *For* $\beta \geq \beta_c$, *the limit*

$$P(\beta) = \lim_{n \to \infty} n^{-1} \log Z_n(\beta)$$

exists almost surely. Moreover, $P(\beta) = \frac{\beta}{\beta_c} \log_c E(W^{\beta_c}) + (\frac{\beta}{\beta_c} - \beta)$.

Concerning the thermodynamic limit of the Gibbs measures for the low temperature case $\beta \geq \beta_c$, it generally fails to exist. By using a compactness argument, the existence of at least one limiting measure is guaranteed; the set of accumulation points of the finite volume Gibbs measures, $\nu_n^\beta(\cdot)$, for an increasing sequence of volumes n is supposed to be non trivial. The study of the topological structure of this set remains an open and important problem for the understanding of the low temperature phase of disordered systems. A plausible conjecture is that exist many (may be infinite) such states. Let me mention in conclusion that some support of this conjecture can be found in the physical theories about of the multiplicity of states for disordered systems and spin glasses.

Acknowledgements. The author wishes to thank the organisers of this workshop for their efforts to bring together mathematicians and physicists working on trees.

References

[1] Biggins, J., Kyprianou, A.: Branching Random Walk: Seneta-Heyde norming, *Trees, Proceedings of a Workshop held in Versailles, June 14–16, 1995* edited by B. Chauvin, S. Cohen, A. Rouault, Birkhäuser (1996).

[2] Chauvin, B., Rouault, A.: Boltzmann-Gibbs weights in the branching random walk. In *Classical and Modern Branching processes*, eds K.B. Athreya, P. Jagers, IMA Proceedings **84**, 41–50. Springer-Verlag, New York.

[3] Collet, P., Koukiou, F.: Large deviations for multiplicative chaos. *Commun. Math. Phys.* **147**, 329–342 (1992).

[4] Guivarc'h, Y.: Sur une extension de la loi semi-stable fonctionnelle non-linéaire de B. Mandelbrot. *Ann. Inst. Henri Poincaré* **26**, 261–285 (1990).

[5] Kahane, J.-P., Peyrière, J.: Sur certaines martingales de Benoit Mandelbrot. *Adv. in Math.* **22**, 131–145 (1976).

[6] Koukiou, F.: The mean field theory of directed polymers in random media and spin glass models. *Rev. in Math. Phys.* **7**, 183–192 (1995).

[7] Mandelbrot, B.: *In Statistical models and turbulence.* Symposium at U.C. San Diego 1971, Springer (1972).

[8] Waymire, E.C., Williams, S.C.: A general decomposition theory for random cascades. *Bulletin of the AMS*, to appear.

[9] Ruelle, D.: *Thermodynamic formalism.* Addison-Wesley (1978).

Flora Koukiou
Milieux Aléatoires et Inhomogènes
Université de Cergy-Pontoise
F-95302 Cergy-Pontoise Cedex

and

Centre de Physique Théorique
Ecole Polytechnique
F-91128 Palaiseau Cedex

koukiou@u-cergy.fr

2. Probability and Trees

Progress in Probability, Vol. 40
© 1996 Birkhäuser Verlag Basel/Switzerland

Branching Random Walk: Seneta-Heyde norming

J.D. BIGGINS[1] AND A.E. KYPRIANOU[2]

Mathematics Subject Classification: 60J80

Keywords: Martingales, functional equations, Seneta-Heyde norming, branching random walk

Abstract. In the discrete-time branching random walk, the martingale formed by taking the Laplace transform of the nth generation point process is known, for suitable values of the argument, to converge in L_1 under an $X \log X$ condition, and to converge to zero when this moment condition fails. This paper examines the strategy used in Biggins and Kyprianou (1996) to prove that, when the $X \log X$ condition fails, there exists a Seneta-Heyde renormalisation of the martingale that converges in probability to a non-trivial random variable. To bring out how the method works it is first discussed in the context of the Galton-Watson process. The paper is concluded by extending the results to the case of the continuous-time Markov branching random walk.

Résumé. Dans une marche aléatoire de branchement à temps discret, on sait que pour des valeurs convenables de l'argument, la martingale obtenue en prenant la transformée de Laplace du processus ponctuel de la n-ième génération converge dans L_1 sous condition $X \log X$ et converge vers zéro quand cette condition de moment n'est pas vérifiée. Ce papier examine la stratégie développée par Biggins et Kyprianou (1996) pour démontrer que lorsque la condition $X \log X$ ne tient plus, il existe une renormalisation de Seneta-Heyde de la martingale qui converge en probabilité vers une variable aléatoire non triviale. Pour éclairer le fonctionnement de la méthode, celle-ci est d'abord discutée dans le cadre d'un processus de Galton-Watson. Le papier se conclut en étendant les résultats au cas d'une marche aléatoire de branchement markovienne à temps continu.

1   Introduction

This paper outlines the strategy used by Biggins and Kyprianou (1996) to prove that martingales arising in the branching random walk have a Seneta-Heyde norming when the appropriate version of an $X \log X$ condition fails. To bring out how the method works it will first be discussed in the context of the Galton-Watson process (though for that case there are several simpler ways to obtain the desired result). Next the case of the discrete-time branching random walk is examined. The paper finishes by extending the results to the continuous-time Markov branching random walk. The reader should note that the emphasis in the discussion of the discrete-time case here is on methodology; rigorous proofs for that case are contained in Biggins and Kyprianou (1996).

---

1)   Some of this work was carried out whilst visiting the Institute of Mathematics and its Applications, University of Minnesota and the Mittag-Leffler Institute, Stockholm. I am grateful to both for their support.

2)   Supported by an EPSRC studentship

As usual, individuals are labelled by their line of descent, so if $u = i_1 \ldots i_n$ then u is the i_nth child of the i_{n-1}th child of ... the i_1th child of the initial ancestor, and $|u|$ is the generation in which u is born. Sums and products involving u will always be over only those that are ever born, but this will not be explicitly stated each time.

Let $Z^{(n)}$ be the number of people in the nth generation, defined recursively by

$$Z^{(n+1)} = \sum_{|u|=n} Z_u,$$

where, given $\mathcal{F}^n$, which is the σ-field describing the ancestry of every individual in the nth generation, $\{Z_u : |u| = n\}$ are independent copies of Z, the family size distribution. In general, a subscript u will be used to indicate quantities associated with u. Let the mean family size, EZ, be $m \in (1, \infty)$. It is well known that $W^{(n)}$, given by $W^{(n)} = m^{-n} Z^{(n)}$, is a (non-negative) martingale, The Kesten-Stigum Theorem, see Asmussen and Hering (1983), shows that this martingale converges in L_1, so that its limit W has mean 1, exactly when $EZ \log Z < \infty$, and the limit is degenerate otherwise. When W is non-degenerate the set $\{W > 0\}$ coincides, almost surely, with the set on which the process survives, that is with $\{Z^{(n)} \to \infty\}$. In general, even when the limit is degenerate, it is possible to find constants $\{l_n\}$ such that

$$l_n W^{(n)} \to \Delta < \infty \quad \text{almost surely,}$$

where $\{\Delta > 0\}$ coincides with the survival set; a result that goes back to Seneta (1968) and Heyde (1970). (Of course when $EZ \log Z < \infty$ the l_n can all be taken to be 1.) There are several ways to prove this result; the neatest is probably that due to Grey (1980), in which an independent copy of the branching process supplies the normalization. However, it does not seem possible to use Grey's method for more complicated branching models.

In the next section a variation of the method employed by Cohn (1982a, 1982b, 1983, 1985) for obtaining a Seneta-Heyde renormalization will be explored in the context of the Galton-Watson process, and the significant steps summarised. The discussion in that section is arranged to lead into, and to facilitate, the corresponding one for the branching random walk, so this presentation of a well-known result is worthwhile. The third section starts with a description of the branching random walk and the martingales which are natural counterparts of $W^{(n)}$. The three key steps, identified during the Galton-Watson discussion, needed to establish a Seneta-Heyde renormalization of these martingales are then examined in turn. The methods used here are strong enough to obtain a Seneta-Heyde renormalization for a general (Crump-Mode-Jagers) branching process. This result, which illustrates that the techniques employed improve on the known results, is discussed briefly in the fourth section.

The main stimulus for the study described in the third and fourth sections was the paper by Neveu (1988) on branching Brownian motion. The fifth section describes briefly how that paper suggested the main result given here. The final section shows how this main result, which is for a discrete-time model, is easily extended to cover the continuous-time Markov case, which includes branching Brownian motion.

2 The Galton-Watson case

The following well-known result is the subject of this section. It only asserts convergence in probability since it is in this form that the result generalizes fully, but, as will become clear, the method actually yields almost sure convergence in the Galton-Watson framework.

Theorem 2.1 *There exist positive constants l_n such that*

$$l_n W^{(n)} \to \Delta \text{ in probability,}$$

as $n \to \infty$, where Δ is finite and strictly positive on the survival set, almost surely.

Note first that, as an elementary consequence of the weak law of large numbers,

$$\frac{W^{(n+1)}}{W^{(n)}} = \frac{Z^{(n+1)}}{mZ^{(n)}} = \frac{1}{Z^{(n)}} \sum_{|u|=n} \frac{Z_u}{m} \to 1 \tag{2.1}$$

in probability on the survival set, as $n \to \infty$.

Let the Laplace transform of $W^{(n)}$ be $\Omega_n(x)$. Take l_n to be such that $\Omega_n(l_n) = \kappa$, where κ is fixed to be greater than the extinction probability but less than one. Choose a subsequence, $\{n(i) : i = 1, 2, \ldots\}$, tending to infinity, such that, along this subsequence, $l_n W^{(n)}$ converges in distribution, with the Laplace transform of the limit being Ψ, so that

$$E \exp(-x l_{n(i)} W^{(n(i))}) \to \Psi(x) \text{ as } i \to \infty.$$

The limit might be an improper variable, but, as a result of the definition of l_n, its mass cannot be concentrated solely at zero or infinity. Thus $\{l_n\}$ is a candidate sequence for the Seneta-Heyde norming. It is worth stressing that, in principle, the limit depends on the subsequence selected.

Let Δ be a variable with transform Ψ. Then, as a consequence of (2.1), Δ must satisfy a distributional equation, and Ψ must satisfy a corresponding functional equation. To see this, note that the branching property implies that each person in the first generation can be viewed as the initial ancestor of an independent copy of the process, so that

$$W^{(n)} = \sum_{|u|=1} m^{-1} W_u^{(n-1)},$$

where, given $\mathcal{F}^1$, $\{W_u^{(n-1)} : |u| = 1\}$ are independent copies of $W^{(n-1)}$. Thus

$$l_n W^{(n)} = \sum_{|u|=1} m^{-1} l_n W_u^{(n)} \frac{W_u^{(n-1)}}{W_u^{(n)}}. \tag{2.2}$$

By taking limits along $\{n(i)\}$, and using (2.1), it follows that

$$\Delta \overset{\mathcal{D}}{=} \sum_{|u|=1} m^{-1}\Delta_u,$$

where $\overset{\mathcal{D}}{=}$ means equality in distribution and, of course, given $\mathcal{F}^1$, $\{\Delta_u : |u| = 1\}$ are independent copies of Δ. Hence, in terms of Ψ,

$$\Psi(x) = E \prod_{|u|=1} \Psi(xm^{-1}).$$

If f is the generating function of the family-size distribution then this equation can be written in the more familiar form

$$\Psi(x) = f(\Psi(xm^{-1})),$$

so that Ψ is a solution to the Poincaré functional equation associated with f. (This argument shows that the existence of a solution to the Poincaré functional equation for a generating function f is a simple consequence of (2.1).)

Letting $x \downarrow 0$ in the Poincaré functional equation shows that $\Psi(0^+)$ satisfies $f(s) = s$, and so must equal either the extinction probability or 1. By arrangement $\Psi(1)$ exceeds the extinction probability, and Ψ is necessarily decreasing so $\Psi(0^+) = 1$; thus Δ is a proper random variable. In a similar way, but letting $x \uparrow \infty$, it follows that Δ has an atom at zero with probability equal to the extinction probability. So, although the variable Δ may depend on the subsequence selected, it is always proper and, once convergence in probability is established, it must be non-zero on the survival set.

The split in (2.2) does not have to be made in the first generation. Splitting on generation k gives (for $n > k$)

$$l_n W^{(n)} = \sum_{|u|=k} m^{-k} l_n W_u^{(n)} \frac{W_u^{(n-k)}}{W_u^{(n)}},$$

so that, taking Laplace transforms conditional on $\mathcal{F}^k$, and using (2.1) again,

$$E[\exp(-x l_{n(i)} W^{(n(i))})|\mathcal{F}^k] \to \prod_{|u|=k} \Psi(xm^{-k}) \quad \text{as } i \to \infty, \tag{2.3}$$

for all positive x, almost surely. The existence of these limiting conditional Laplace transforms and their properties allow the results described next to be brought into play.

Suppose that $\{Y_n\}$ is a sequence of non-negative random variables adapted to the increasing σ-fields $\{\mathcal{G}^n\}$ and that along a fixed subsequence $\{n(i) : i = 1, 2, \ldots\}$, for each k and $x > 0$, the conditional Laplace transform $E[\exp(-xY_{n(i)})|\mathcal{G}^k]$ converges as $i \to \infty$. Denote the limit by $\psi_k(x)$. The first lemma is a simple exercise in conditional expectations.

Lemma 2.2 *For each $x > 0$, $\{\psi_k(x)\}$ forms a bounded non-negative martingale with respect to $\{\mathcal{G}^k\}$.*

The martingale $\{\psi_k(x)\}$ converges almost surely and in mean to a limit which will be denoted by $\psi(x)$. It turns out that when this limit has a suitable simple form the convergence in distribution of $\{Y_{n(i)} : i\}$ can be strengthened to convergence in probability. This is the content of the next lemma.

Lemma 2.3 *If, for $x > 0$, $\psi(x) = e^{-xX}$ for a finite random variable X (that does not depend on x) then $Y_{n(i)} \to X$ in probability, as $i \to \infty$.*

To explain the idea of the proof let $X_k = -\log \psi_k(1)$, so that the convergence of the martingale $\{\psi_k(1)\}$ implies that $X_k \to X$ almost surely. The convergence in probability is proved by showing that, for a suitable sequence $\{k(i)\}$, $Y_{n(i)} - X_{k(i)}$ converges in distribution to zero. To see why this should work, compute the appropriate Laplace transform:

$$\begin{aligned}
E[\exp(-x(Y_{n(i)} - X_{k(i)}))] &= E[E[\exp(-x(Y_{n(i)} - X_{k(i)}))|\mathcal{G}^{k(i)}]] \\
&= E[\exp(xX_{k(i)})E[\exp(-xY_{n(i)})|\mathcal{G}^{k(i)}]] \\
&\approx E[\exp(xX_{k(i)})\psi_{k(i)}(x)] \qquad \text{(large } n(i)) \\
&\approx E[\exp(xX)\exp(-xX)] = 1 \quad \text{(large } k(i)),
\end{aligned}$$

where it is the last step that uses the special form of the limit of the martingales $\psi_k(x)$. Convergence in distribution to a constant implies convergence in probability, so a rigorous version of this argument will provide a proof of the Lemma.

These lemmas can be compared with Theorem 3.1 in Cohn (1985), which plays a similar role in that study. That theorem is phrased in terms of random variables rather than Laplace transforms; the formulation adopted here is better suited to the present approach.

Applying Lemma 2.2 to (2.3) yields that, for each $x > 0$, $\prod_{|u|=k} \Psi(xm^{-k})$ is a martingale. Denote the corresponding limits by $M(x)$. At this point in the Galton-Watson case it is quite simple to finish off the argument for Seneta-Heyde norming with almost sure convergence, because this martingale is a simple function of the variables for which a renormalization is sought. Thus

$$\prod_{|u|=k} \Psi(xm^{-k}) = \Psi(xm^{-k})^{Z^{(k)}} = \exp(-Z^{(k)}(-\log \Psi(xm^{-k}))) \to M(x)$$

as $k \to \infty$, so that, for any $x > 0$, $\{-m^k \log \Psi(xm^{-k})\}$ will serve as $\{l_k\}$ and then

$$l_k W^{(k)} \to -\log M(x) \quad \text{almost surely.}$$

Of course different normalizations, arising from different values of x, can only give rise to scalar multiples of a single limit.

Ignoring this simplification, hoping instead to use Lemma 2.3, focuses attention on how the limits $\log M(x)$ depend on x. Note that

$$-\log M(x) = -\log\left(\lim_k \prod_{|u|=k} \Psi(xm^{-k})\right) = \lim_k \sum_{|u|=k} -\log\Psi(xm^{-k})$$

$$\approx \lim_k \sum_{|u|=k} (1 - \Psi(xm^{-k}))$$

so, when $(1 - \Psi(x))/x$ is slowly varying at zero,

$$-\log M(x) \approx \lim_k \sum_{|u|=k} x(1 - \Psi(m^{-k})) \approx -x\log M(1).$$

This indicates that establishing that $(1 - \Psi(x))/x$ is slowly varying is exactly the issue in showing that the martingale limits $M(x)$ have the right form for Lemma 2.3 to apply. (For the Poincaré functional equation it is a simple exercise in analysis to show that, for a solution that is a Laplace transform, the required slow variation obtains.) In this way convergence in probability along the selected subsequence is established. The proof is now completed by using the fact that the solution to the Poincaré functional equation is unique, up to a scale factor, so that all subsequences have the same limit. Uniqueness, based on analytical results about the functional equation, is noted in, for example, Lemma 4.1 and Theorem 4.2 of Seneta(1969).

It is worth identifying explicitly the main steps involved in finding the renormalization of $W^{(n)}$.

Step 1:   **A law of large numbers.** Show that $W^{(n+1)}/W^{(n)}$ converges to 1 in probability. Select candidate norming constants $\{l_n\}$ that give convergence in distribution of $\{l_n W^{(n)}\}$ along a subsequence. Now use the branching property to show two things: the limit of the renormalized subsequence satisfies a distributional equation, so its transform, Ψ, satisfies the corresponding functional equation; and the limit of the conditional Laplace transforms of the renormalized sequence exists, and so provides martingales with associated non-trivial limits by Lemma 2.2.

Step 2:   **Slow variation.** Show, by studying the functional equation, that $(1 - \Psi(x))/x$ is slowly varying. Use this property to show that the martingale limits arising from Lemma 2.2 have the properties necessary for Lemma 2.3 to apply, thereby yielding convergence in probability of $\{l_n W^{(n)}\}$ along a subsequence.

Step 3:   **Uniqueness.** Establish the uniqueness of the solution to the functional equation and use it to show that all subsequences have the same limit, so that convergence holds along the full sequence.

3   The branching random walk

The notation used for the branching random walk will subsume that already in use for the Galton-Watson process. An initial ancestor is at the origin of the real line, and the positions of her children are given by a point process Z. Each of these children has children in the same way, in that the positions of each family are, relative to the parent, given by an independent copy of Z, and so on. Let $Z^{(n)}$ be the point process formed by the nth generation, with points $\{z_u : |u| = n\}$; then, by definition, for any set A

$$Z^{(n+1)}(A) = \sum_{|u|=n} Z_u\left(A - z_u\right), \tag{3.1}$$

where, given $\mathcal{F}^n$, $\{Z_u : |u| = n\}$ are independent copies of Z.

Suppose Z has intensity measure μ with Laplace transform m, so that

$$m\left(\phi\right) = E\left[\sum_{|u|=1} \exp(-\phi z_u)\right].$$

Throughout it will be assumed that: $m(\phi)$ is finite on some interval; θ is a value within this interval; the process is supercritical, so that $m(0) > 1$; and family sizes are finite. For convenience, it will also be assumed that $\theta \geq 0$; this is no loss of generality as a reflection of the positions about the origin shows.

It is well known, and easily shown, that

$$W^{(n)}(\theta) = m\left(\theta\right)^{-n} \int e^{-\theta x} Z^{(n)}(dx) = \sum_{|u|=n} \frac{\exp(-\theta z_u)}{m\left(\theta\right)^n}$$

is a martingale with respect to the σ-fields $\{\mathcal{F}^n\}$. This martingale is positive and so has an almost sure limit $W(\theta)$ which, by Fatou's lemma, satisfies $E[W(\theta)] \leq 1$. When $\theta = 0$ this martingale is the Galton-Watson one discussed in the previous section.

The convergence in mean of the martingale $W^{(n)}(\theta)$ or similar martingales arising from slightly different processes is considered in, for example, Kingman (1975), Kahane and Peyrière (1976), Biggins (1977a), Neveu (1988), Lyons (1995). In particular, the next result, which is an analogue of the Kesten-Stigum Theorem, is a consequence of those in Biggins (1977a).

Theorem 3.1 *The martingale $W^{(n)}(\theta)$ converges in L_1, so that $EW(\theta) = 1$, if and only if*

$$\log m(\theta) - \theta m'(\theta)/m(\theta) > 0 \tag{3.2}$$

and

$$E[W^{(1)}(\theta) \log^+ W^{(1)}(\theta)] < \infty, \tag{3.3}$$

and $EW(\theta) = 0$ otherwise.

It is worth noting that the set of θ values satisfying (3.2) intersect with the interior of the domain of finiteness of m to form an open interval.

In the light of this theorem, it is natural to seek a Seneta-Heyde norming for the martingale $W^{(n)}(\theta)$, that is, to prove the following theorem.

Theorem 3.2 *When (3.2) holds there exists a sequence of constants $\{l_n\}$ such that*

$$l_n W^{(n)}(\theta) \to \Delta \quad \text{in probability},$$

as $n \to \infty$, where Δ is a finite random variable which is strictly positive when the process survives. (In general, both $\{l_n\}$ and Δ depend on θ.)

Since θ is fixed it will be omitted in the notation whenever possible. To simplify the notation further, let

$$y_u = \frac{\exp(-\theta z_u)}{m(\theta)^{|u|}}.$$

Thus, with these conventions, $W^{(n)} = \sum_{|u|=n} y_u$. The issues that arise in following through the method outlined in the previous section will now be examined.

Step 1: A law of large numbers

By looking at the branching processes stemming from each kth generation person it is easily seen that, for $n > k$,

$$W^{(n)} = \sum_{|u|=k} y_u W_u^{(n-k)}, \tag{3.4}$$

where, given $\mathcal{F}^k$, $\{W_u^{(n-k)} : |u| = k\}$ are independent copies of $W^{(n-k)}$. Thus

$$\frac{W^{(n+1)}}{W^{(n)}} = \sum_{|u|=n} \frac{y_u}{W^{(n)}} W_u^{(1)}$$

and so is, given $\mathcal{F}^n$, a weighted sum of independent identically distributed variables. Kurtz (1972) provides tools for studying the convergence of such weighted sums; using these leads to the following analogue of (2.1).

Theorem 3.3 *When (3.2) holds*

$$\frac{W^{(n+1)}}{W^{(n)}} \to 1, \quad \text{in probability}, \tag{3.5}$$

as $n \to \infty$, on the survival set of the process.

The norming constants are defined through the Laplace transform of $W^{(n)}$, just as in the Galton-Watson case, and a suitable subsequence of $l_n W^{(n)}$ converges in distribution. The limit will again be denoted by Δ, with transform Ψ. Then, using (3.4) and Theorem 3.3, Δ satisfies the distributional equation

$$\Delta \stackrel{\mathcal{D}}{=} \sum_{|u|=1} y_u \Delta_u,$$

where, given $\{y_u : |u| = 1\}$, Δ_u are independent copies of Δ. In terms of transforms this becomes

$$\Psi(x) = E\left[\prod_{|u|=1} \Psi(xy_u)\right]. \tag{3.6}$$

This functional equation, or variants of it, has been much studied, see, for example, Kahane and Peyrière (1976), Biggins (1977a), Durrett and Liggett (1983), Pakes (1992), Liu (1995). The argument that Δ is proper and takes the value zero with the extinction probability is just as in the Galton-Watson case. Furthermore, using (3.4) and Theorem 3.3 again, as n goes to infinity along the selected subsequence

$$E[\exp(-xl_nW^{(n)})|\mathcal{F}^k] \to \prod_{|u|=k} \Psi(xy_u) =: M^{(k)}(x), \tag{3.7}$$

for all positive x, almost surely. By Lemma 2.2, the right hand side here is a bounded martingale; its limit is denoted by $M(x)$.

The martingale $W^{(n)}$ is obtained by adding contributions from each nth generation person, and so may reasonably be called an additive martingale. In contrast the martingales $M^{(n)}$ result from multiplying terms; it is natural therefore to call these multiplicative martingales. This is the terminology used by Neveu (1988) in his study of similar martingales arising in the context of branching Brownian motion.

Step 2: Slow variation

Theorem 3.4 *When (3.2) holds any solution Ψ to the functional equation (3.6) that is the Laplace transform of a non-trivial random variable is such that $L(x) := x^{-1}(1 - \Psi(x))$ is slowly varying as $x \downarrow 0$.*

Under the additional condition that $m(0) < \infty$, this theorem is a consequence of Theorem 2 of Liu (1995). In the context of the Bellman-Harris process, which gives rise to a particular case of equation (3.6), Theorem 3.4 follows from Schuh (1982), as is pointed out in Cohn (1983). The general, or Crump-Mode-Jagers, branching process gives rise to a functional equation just like (3.6), save only that, for all u, $y_u < 1$. Slow variation for this case is stated in Theorem 6.2 of Cohn (1985).

Provided $\sup_{|u|=k} y_u \downarrow 0$ (which does hold), Theorem 3.4 allows us to make the second of the following approximations as k gets large,

$$-\log M^{(k)}(x) \approx \sum_{|u|=k} (1 - \Psi(xy_u))$$

$$\approx \sum_{|u|=k} x(1 - \Psi(y_u)) \approx -x \log M^{(k)}(1).$$

Thus, in the limit as $k \to \infty$,

$$-\log M(x) = -x \log M(1), \tag{3.8}$$

and Lemma 2.3 applies to give that, along the selected subsequence, $\{l_n W^{(n)}\}$ converges in probability to $-\log M(1)$. This implies, in particular, that Δ is $-\log M(1)$.

Note too that

$$-\log M^{(n)}(1) \approx \sum_{|u|=n} (1 - \Psi(y_u)) = \sum_{|u|=n} L(y_u)y_u$$

so that when $W^{(n)}$ converges in mean, which implies that $L(0^+) = 1$,

$$-\log M^{(n)}(1) \approx \sum_{|u|=n} y_u = W^{(n)};$$

thus the logarithm of the multiplicative martingale and the additive martingale are then asymptotically equivalent.

Step 3: Uniqueness

Theorem 3.5 *When* (3.2) *holds, the non-trivial solution to the functional equation* (3.6) *is unique, within the class of Laplace transforms of non-negative variables (up to a multiplicative constant in the argument).*

It is this theorem that provided the greatest difficulties in establishing the main result. It is easy to see that any solution to the functional equation has a multiplicative martingale of the form introduced earlier (the right side of (3.7)) associated with it. Furthermore, as the martingale converges in mean, $EM(x) = \Psi(x)$, which combines with (3.8) to show that Ψ is the Laplace transform of $-\log M(1)$.

One tempting way forward is to try to use the slow variation of L and the approximation $-\log M^{(n)}(1) \approx \sum_{|u|=n} L(y_u)y_u$, mentioned at the end of Step 2, to show that

$$-\log M^{(n)}(1) \approx \sum_{|u|=n} L(y_u)y_u \approx L(a_n) \sum_{|u|=n} y_u = L(a_n)W^{(n)}$$

for some deterministic sequence a_n. Then the almost sure convergence of the multiplicative martingale would imply the almost sure convergence of $L(a_n)W^{(n)}$, which is a stronger result than that claimed in Theorem 3.2. Furthermore, much as in the uniqueness proof for the Poincaré equation given in Theorem III.5.2 in Asmussen and Hering (1983), two supposedly different solutions to the functional equation, Ψ_1 and Ψ_2, with associated multiplicative martingale limits M_1 and M_2, and sequences $a_n^{(1)}$ and $a_n^{(2)}$ respectively, then give

$$\frac{-\log M_1(1)}{-\log M_2(1)} = \lim_n \frac{L_1(a_n^{(1)})W^{(n)}}{L_2(a_n^{(2)})W^{(n)}} = \lim_n \frac{L_1(a_n^{(1)})}{L_2(a_n^{(2)})}.$$

The right hand limit thus exists and must be equal to some constant in $(0, \infty)$ because $-\log M_1(1)$ and $-\log M_2(1)$ are strictly positive on the survival set. Hence M_1 is a scalar multiple of M_2, so that Ψ_1 and Ψ_2 differ only by a scale factor, giving uniqueness.

Unfortunately the vital step in this line of reasoning, that the terms $\{L(y_u) : |u| = n\}$ can all be well approximated by $L(a_n)$ for some a_n because L is slowly varying, fails because the values of $\{y_u : |u| = n\}$ are too disperse. (It is easy to use the results in Biggins (1977b) to see that the ratio between the smallest and largest term in $\{y_u : |u| = n\}$ grows geometrically in n.) The way round this problem is to construct new martingales that are better behaved in this respect.

Let $\mathcal{I}(s)$ be the set of individuals who are the first in their line of descent to have y_u less than e^{-s}, so

$$\mathcal{I}(s) = \{u : y_u < e^{-s}, \text{ but } y_v \geq e^{-s} \text{ for } v < u\},$$

where $v < u$ if v is a strict ancestor of u. Note that

$$\mathcal{I}(s) = \{u : \theta z_u + |u| \log m(\theta) > s, \text{ but } \theta z_v + |v| \log m(\theta) \leq s \text{ for } v < u\},$$

so that this set of individuals are the first in their lines of descent to cross the space-time line $\theta p + t \log m(\theta) = s$, where (p, t) are (space,time) coordinates.

The sets $\{\mathcal{I}(s) : s\}$ form a collection that is totally ordered as s varies, in the sense that all members of any of these sets have ancestors in any earlier one. The collection $\{u : |u| = n\}$ is similarly ordered as n varies. By developing a suitable general theory, most of which already exists in the work of Chauvin (1988, 1991) and Jagers (1989), it can be shown that

$$M^{(\mathcal{I}(s))}(x) := \prod_{u \in \mathcal{I}(s)} \Psi(x y_u)$$

is a martingale with the same limit as the martingale $M^{(n)}(x)$; thus

$$M^{(\mathcal{I}(s))}(x) \to M(x) \text{ almost surely, as } s \to \infty.$$

The martingales $M^{(\mathcal{I}(s))}$ and $M^{(n)}$ have the same form, differing only in the individuals that the products are taken over. Taking the product over $\mathcal{I}(s)$ offers the big advantage that it ensures that, by arrangement, the terms in $\{y_u : u \in \mathcal{I}(s)\}$ are all quite near to e^{-s}. Thus the argument mentioned previously, where we replace $L(y_u)$ by something deterministic, has a much better chance of working when applied to this martingale.

To see what more is required suppose first that when a line of descent crosses a space-time line the exceedance is bounded. That is, in terms of the basic point process, the set $\{z_u : |u| = 1\}$ is bounded above, so that, for some c, $\sup\{\theta z_u +$

$\log m(\theta) : |u| = 1\} < c$. Expressing this in terms of y_u for $u \in \mathcal{I}(s)$, there is a c such that $e^{-(s+c)} < y_u < e^{-s}$. Thus, as L is monotone,

$$L(e^{-s}) \sum_{u \in \mathcal{I}(s)} y_u \leq \sum_{u \in \mathcal{I}(s)} L(y_u) y_u \leq L(e^{-(s+c)}) \sum_{u \in \mathcal{I}(s)} y_u,$$

and the slow variation of L, and a little analysis, yields the almost sure asymptotic equivalence

$$L(e^{-s}) \sum_{u \in \mathcal{I}(s)} y_u \sim - \sum_{u \in \mathcal{I}(s)} \log \Psi(y_u) \quad \text{as } s \to \infty; \tag{3.9}$$

the second expression is the logarithm of the multiplicative martingale $M^{(\mathcal{I}(s))}(1)$, which has, as has already been noted, the non-trivial limit $-\log M(1)$. The argument for uniqueness suggested earlier works without further problems in this case.

To cover the general case an argument is needed to show that the possibility of arbitrarily large exceedance over the space-time line does not disturb the limiting behaviour. The approach used is based on results obtained by Nerman (1981). Suppose we define $\mathcal{I}(s, c)$ to be those individuals in $\mathcal{I}(s)$ but with an exceedance over the line greater than c. Then the argument leading to the equivalence in (3.9) applies if sums are taken over $u \in \mathcal{I}(s) \setminus \mathcal{I}(s, c)$. The issue then becomes to show that the part omitted can be made as small as desired or, more precisely, after some straightforward manipulations, to show that

$$\lim_{c \to \infty} \lim_{t \to \infty} \frac{\sum_{\mathcal{I}(t,c)} e^t y_u L(y_u)/L(e^{-t})}{\sum_{\mathcal{I}(t) \setminus \mathcal{I}(t,c)} e^t y_u} = 0. \tag{3.10}$$

As slowly varying functions grow more slowly than any power, it will, in place of (3.10) be enough to show that, for sufficiently small $\epsilon > 0$,

$$\lim_{c \to \infty} \lim_{t \to \infty} \frac{\sum_{\mathcal{I}(t,c)} (e^t y_u)^{1-\epsilon}}{\sum_{\mathcal{I}(t) \setminus \mathcal{I}(t,c)} e^t y_u} = 0. \tag{3.11}$$

The proof now proceeds by identifying an embedded general branching process, with birth times $\{-\log y_u\}$, thereby allowing Theorem 6.3 of Nerman (1981), on the convergence of the ratios of a general branching process counted by two different characteristics, to be invoked to prove (3.11) and hence the uniqueness claimed in Theorem 3.5.

Both Theorem 6.3 of Nerman (1981) and (3.9) with sums taken over $u \in \mathcal{I}(s) \setminus \mathcal{I}(s, c)$, are results that hold almost surely, so the proof just outlined shows that the asymptotic equivalence in (3.9) holds almost surely in the general case. This is a Seneta-Heyde renormalization of the additive martingale $\sum_{u \in \mathcal{I}(s)} y_u$ that converges almost surely. In contrast, the renormalization of the original additive martingale $W^{(n)} (= \sum_{|u|=n} y_u)$ is proved to converge using Lemma 2.3, which yields only convergence in probability.

It is worth giving a little more detail on how the embedded general branching process that is at the heart of the proof of uniqueness arises. Note first that $\mathcal{I}(0)$ is the set of individuals that are the first in their line of descent to have $-\log y_u > 0$. To construct a suitable embedded process, regard these individuals as the children of the initial ancestor, with birth times $\{-\log y_u : u \in \mathcal{I}(0)\}$. Of course these individuals need not be in the first generation of the original process and so need not be true children of the initial ancestor. Consider the space-time lines $\theta p + t \log m(\theta) = s$ as s varies. When $s = 0$ the collection $\mathcal{I}(0)$ are the individuals who are the first in their line of descent to be on the positive side of the line. This is illustrated in Figure 1 where the individuals coloured black belong to the set $\mathcal{I}(0)$. (Note that this figure shows a sample path of only the first six generations of some branching random walk).

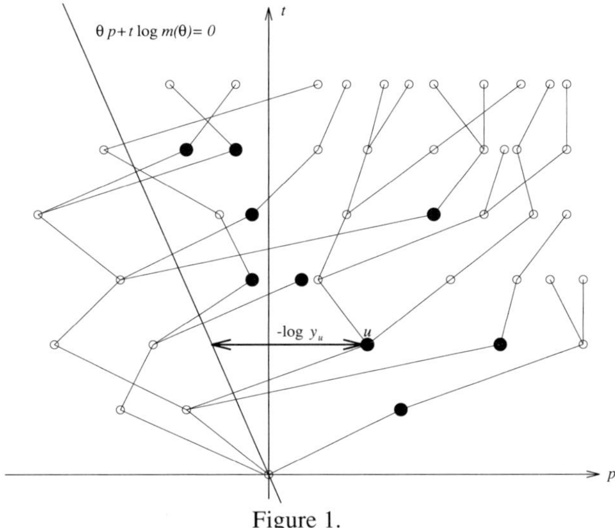

Figure 1.

As s increases the space-time line sweeps past individuals. Suppose that at s', the line passes through an individual $u \in \mathcal{I}(0)$, so that, for any $s < s', u \in \mathcal{I}(s)$ but $u \notin \mathcal{I}(s')$. As u drops out of $\mathcal{I}(s)$ it is replaced by a collection of its descendants; this is formed by the copy of $\mathcal{I}(0)$ associated with the descendants of u. It is these individuals that are to be regarded as the children of u in the next (embedded) generation. An example of this can be seen in Figure 2. The darkened branches indicate the tree generated by u and the individuals in this tree coloured black make up u's copy of $\mathcal{I}(0)$. In this way it is shown that $\{-\log y_u : u \in \mathcal{I}(s)$ for some $s > 0\}$ are the birth times for a general (Crump-Mode-Jagers) branching process, and $\mathcal{I}(s)$ is what is, in that context, often called the coming generation, which is, at time s, the individuals not yet born but whose mothers are. See Figure 3 in which individuals coloured black are members of the coming generation, $\mathcal{I}(s)$.

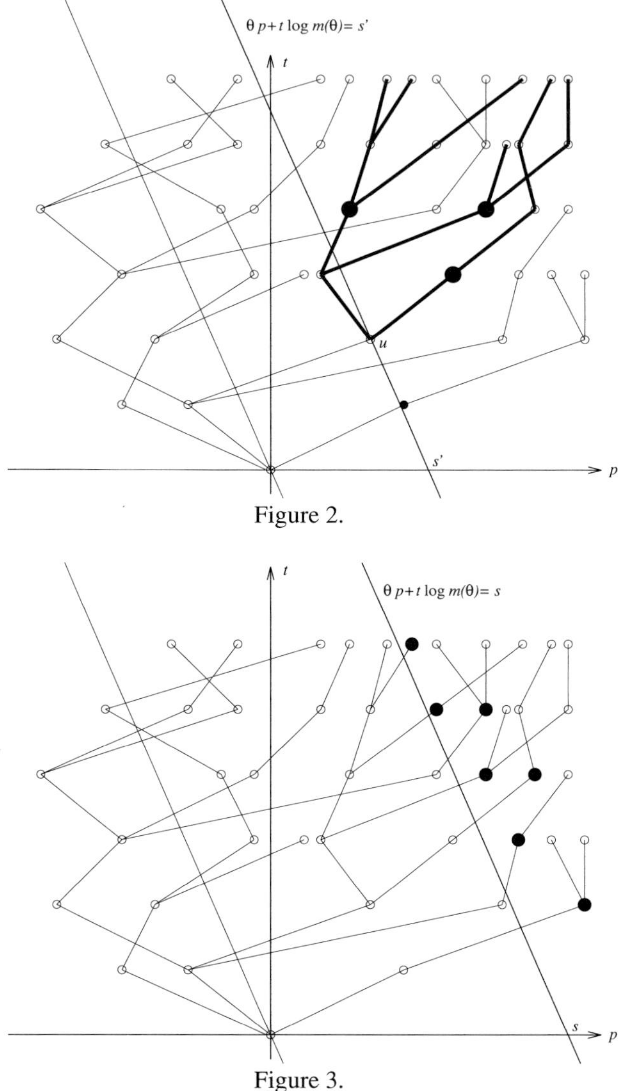

Figure 2.

Figure 3.

4   Seneta-Heyde norming in the general branching process

By identifying position and birth-time, any general branching process can be treated as a branching random walk with the special feature that children are always born to the right of the parent. In this special case the embedded general branching process described in the previous section is the whole process. Then (3.9) gives the Seneta-Heyde normalization of Nerman's martingale, see Nerman (1981), with almost sure convergence. As Cohn (1985) points out, this is sufficient, because of Theorem 6.3 of Nerman (1981), to establish the Seneta-Heyde norming for any way of counting the process, providing the counting characteristic ϕ is sufficiently well behaved. (Jagers (1975) can be consulted for a discussion of the notion of characteristics.) Expressing this accurately gives the following result.

Theorem 4.1 *Consider a general (C-M-J) branching process with finite family sizes, reproduction intensity measure η, Malthusian parameter $\alpha > 0$, and with birth times $\{b_u\}$. Suppose that there is a $\beta \in [0, \alpha)$ such that $\int e^{-\beta\sigma}\eta(d\sigma) < \infty$. Then there is a slowly varying function L such that, for any characteristic $\phi(t)$ (with D-paths) and $E(\sup_t e^{-\beta t}\phi(t)) < \infty$,*

$$L(e^{-\alpha t}) \sum_u \phi_u(t - b_u)$$

has an almost sure limit as $t \to \infty$. The limit is finite, and non-zero when the process survives.

This is an improvement on the result obtained by Cohn (1985) in that he assumed a finite mean family size, so that η is a finite measure, which is not needed in the approach outlined here. As mentioned in the Introduction, Theorem 4.1 illustrates that, though Theorem 3.2 gives only convergence in probability, the study yields, as a by-product, almost sure convergence, under improved conditions, in the nearest previously considered problem. This suggests that to strengthen the convergence in Theorem 3.2, if such a thing is possible, some new idea will be needed.

5 Branching Brownian motion

The purpose of this section is to outline the results that were the main stimulus for embarking on the developments described in the previous two sections. It starts by giving an account of a few of the results described in Neveu (1988).

Consider a branching Brownian motion with binary splitting starting from a single particle at the origin. Let $Z^{(t)}$ be the point process giving the occupied positions at time t. For $a \geq \sqrt{2}$, the differential equation

$$\frac{1}{2}\Phi'' + a\Phi' + (\Phi^2 - \Phi) = 0 \quad \text{with } \Phi(-\infty) = 0, \ \Phi(\infty) = 1, \tag{5.1}$$

has a unique solution, up to a translation. (Note that the more usual formulation, used in Neveu (1988), reflects the problem about the origin.) Then, for any x,

$$\exp\left(\int \log \Phi(at - x - p)Z^{(t)}(dp)\right) \tag{5.2}$$

forms a (multiplicative) martingale. For $a > \sqrt{2}$, as $x \to \infty$,

$$1 - \Phi(x) \sim ce^{bx} \quad \text{for suitable } b \text{ and } c. \tag{5.3}$$

This estimate implies, again for $a > \sqrt{2}$, that the logarithm of the martingale (5.2) is asymptotically equivalent to the (additive) martingale,

$$\int e^{-b(p-ta)} Z^{(t)}(dp). \tag{5.4}$$

Though the notation is different this is the same equivalence noted at the end of Step 2 in Section 3.

The product in (5.2) is over all particles present at time t. This can be regarded as a product over the positions where lines of descent first hit this time line. Neveu makes the pretty observation that if, instead, the product is taken over the positions where lines of descent first meet a suitable space-time line, specifically (p, t) such that $s = at - p$, all terms in the product are equal, and, as parallel space-time lines are considered, that is, as s changes, the number of terms in the product develops as a Markov branching process. (The theory needed to make this observation rigorous is given by Chauvin (1991).) This Markov branching process is the counterpart of the embedded general branching process that plays an important part in Step 3 in Section 3. As Brownian paths are continuous, overshoot of space-time lines is impossible, making some arguments easier.

Once the assumption of binary splitting is abandoned, in which case Φ^2 in (5.1) is replaced by $f(\Phi)$, where f is the family-size generating function, the equivalence Neveu noted between the additive and multiplicative martingales can break down. Assume the mean family size is finite. Then, because the differential equation has a solution, the multiplicative martingales exist and have non-trivial limits. However the continuous analogue of Theorem 3.1 shows that when the family size does not have a finite $X \log X$ moment the additive martingale converges to zero. Theorem 2.1 in Uchiyama (1978) can be consulted to see that, once $X \log X$ fails, the vital property (5.3) no longer holds. (Uchiyama's condition is, effectively, in terms of f, but its equivalence to the moment condition is standard.) Analogy with the Galton-Watson case suggested that when $X \log X$ fails the limit of a multiplicative martingale should provide the non-trivial limit for a suitable renormalization of the additive martingale. This study shows that this is indeed the case.

6   Continuous-time BRW

It is fairly easy, using a skeleton argument, to extend the discrete-time result given in Theorem 3.2 to the continuous-time Markov case, which includes branching Brownian motion discussed in the previous section. The initial ancestor lives for an exponential length of time, during which she moves according to a process with independent increments (with D-paths), at the end of this time she divides into children which are scattered about her present position according to a point process, X. Each of these children then lives, moves and divides in a similar way to the initial ancestor, independently of each other, and so on. In branching Brownian motion the movement is Brownian and X attaches weight only to the origin.

Let $Z^{(t)}$ be the point process at time t, and let $m(\phi) = E \int e^{-\phi x} Z^{(1)}(dx)$. The process obtained by sampling the continuous-time process only at integer times is a discrete-time branching random walk, so this definition of m fits with the earlier one. The Laplace transform m will be assumed to satisfy the same conditions as in the discrete case. Then

$$W^{(t)}(\theta) = m(\theta)^{-t} \int e^{-\theta x} Z^{(t)}(dx)$$

is a continuous-time martingale, which in branching Brownian motion is the martingale (5.4) in another notation. A discussion of the L_1 convergence of this martingale and of its convergence as a function of θ can be found in the final section of Biggins (1992).

Assume (3.2) holds. Then, by Theorem 3.2, which applies to the skeleton, there exists a sequence of constants $\{l_n\}$ such that $l_n W^{(n)}$ converges in probability to Δ. Let $[t]$ be the largest integer smaller than t and let $l_t = l_{[t]+1}$. It will now be shown that $l_t W^{(t)}$ also converges in probability to Δ.

Note first that, because $l_n W^{(n)}$ converges, for any finite x,

$$(l_{n+1} W^{(n+1)} - l_n W^{(n)}) I(l_n W^{(n)} \leq x) \to 0 \text{ in probability.}$$

Using dominated convergence, this can be written in terms of Laplace transforms as

$$E[\exp(-u(l_{n+1} W^{(n+1)} - l_n W^{(n)}) I(l_n W^{(n)} \leq x))] \to 1.$$

As $W^{(t)}$ is a martingale, two applications of Jensen's inequality give, for $n < t < n+1$,

$$E[\exp(-u(l_{n+1} W^{(n+1)} - l_n W^{(n)}) I(l_n W^{(n)} \leq x))]$$
$$= E[E[\exp(-u(l_{n+1} W^{(n+1)} - l_n W^{(n)}) I(l_n W^{(n)} \leq x)) | \mathcal{F}^t]]$$
$$\geq E[E[\exp(-u(l_{n+1} W^{(t)} - l_n W^{(n)}) I(l_n W^{(n)} \leq x)) | \mathcal{F}^n]]$$
$$\geq E[\exp(-u(l_{n+1} W^{(n)} - l_n W^{(n)}) I(l_n W^{(n)} \leq x))]$$
$$\to 1 \text{ as } n \to \infty.$$

Thus

$$(l_t W^{(t)} - l_n W^{(n)}) I(l_n W^{(n)} \leq x))$$

converges in probability to 0 as $t \to \infty$. Now

$$|l_t W^{(t)} - \Delta| \leq |l_n W^{(n)} - \Delta| + |(l_t W^{(t)} - l_n W^{(n)}) I(l_n W^{(n)} \leq x)|$$
$$+ |(l_t W^{(t)} - l_n W^{(n)}) I(l_n W^{(n)} > x)|;$$

the first two terms converge to zero in probability, whilst the final one is non zero with a probability close to $P(\Delta > x)$ for large n. As Δ is finite and x is arbitrary this last probability can be made as small as desired, thereby showing that, in probability,

$$l_t W^{(t)} \to \Delta.$$

When branching Brownian motion is approached in this way the functional equation (3.6) for the first generation of the discrete skeleton is used to introduce the multiplicative martingales, rather than the differential equation

$$\frac{1}{2}\Phi'' + a\Phi' + (f(\Phi) - \Phi) = 0 \quad \text{with } \Phi(-\infty) = 0, \ \Phi(\infty) = 1. \qquad (6.1).$$

If Ψ solves the functional equation it is easy to show that it also solves the equation

$$\Psi(x) = E \exp \int \log \Psi \left(x \frac{e^{-\theta p}}{m(\theta)^t} \right) Z^{(t)}(dp).$$

It is then straightforward to show, following Appendix A of Bramson (1978), that, writing $\Phi(x) := \Psi(e^{-\theta x})$,

$$u(x,t) = E \exp \int \log \Phi \, (x + p) \, Z^{(t)}(dp) = \Phi \left(x - \frac{\log m(\theta)}{\theta} t \right)$$

solves the KPP equation $u_t = (1/2)u_{xx} + f(u) - u$ and consequently Φ solves (6.1) with $a = \theta^{-1} \log m(\theta)$.

References

[1]   Asmussen, S. and Hering, H. (1983). *Branching Processes*. Birkhäuser, Boston.

[2]   Biggins, J.D. (1977a). Martingale convergence in the branching random walk. *J. Appl. Probab.* **14**, 25–37.

[3]   Biggins, J.D. (1977b). Chernoff's Theorem in the branching random walk. *J. Appl. Probab.* **14**, 630–636.

[4]   Biggins, J.D. (1992). Uniform convergence of martingales in the branching random walk. *Ann. Probab.* **20**, 137–151.

[5]   Biggins, J.D. and Kyprianou, A.E. (1996). Seneta-Heyde norming in the branching random walk. *Ann. Probab.* To appear.

[6]   Bramson, M. (1979). Maximal displacement of branching Brownian motion. *Comm. Pure Appl. Math.* **31**, 531–581.

[7]   Chauvin, B. (1988). *Arbres et Processus de Branchement*. These de Doctorat, Université Paris 6.

[8]   Chauvin, B. (1991). Product martingales and stopping lines for branching Brownian motion. *Ann. Probab.* **19**, 1195–1205.

[9]   Cohn, H. (1982a). Another look at the finite mean supercritical Bienaymé-Galton-Watson process. *J. Appl. Probab.* **19A**, 307–312.

[10] Cohn, H. (1982b). Norming constants for the finite mean supercritical Bellman-Harris process. *Z. Wahrsch. verw. Gebiete* **61**, 189–205.

[11] Cohn, H. (1983). On the convergence result for the supercritical Bellman-Harris process. *Aust. J. Statist.* **25**, 249–255.

[12] Cohn, H. (1985). A martingale approach to supercritical (CMJ) branching processes. *Ann. Probab.* **13**, 1179–1191.

[13] Grey, D.R. (1980). A new look at convergence of branching processes. *Ann. Probab.* **8**, 377–380.

[14] Heyde, C.C. (1970). Extension of a result of Seneta for the supercritical branching process. *Ann. Math. Statist.* **41**, 739–742.

[15] Jagers, P. (1989). General branching processes as Markov fields. *Stoc. Proc. Appl.* **32**, 183–212.

[16] Kahane, J.P. and Peyrière, J. (1976). Sur certaines martingales de Benoit Mandelbrot. *Adv. Math.* **22**, 131–145.

[17] Kingman, J.F.C. (1975). The first birth problem for an age-dependent branching process. *Ann. Probab.* **3**, 790–801.

[18] Kurtz, T.G. (1972). Inequalities for the law of large numbers. *Ann. Math. Statist.* **43**, 1874–1883.

[19] Liu, Q. (1995). Fixed points of a generalized smoothing transformation and applications to branching processes. *J. Appl. Probab.* Submitted.

[20] Lyons, R. (1996). A simple path to Biggins' martingale convergence. In *Classical and Modern Branching processes*, eds K.B. Athreya, P. Jagers, IMA Proceedings **84**, 217–222. Springer-Verlag, New York.

[21] Nerman, O. (1981). On the convergence of supercritical general (C-M-J) branching process. *Z. Wahrsch. verw. Gebiete* **57**, 365–395.

[22] Neveu, J. (1988). Multiplicative martingales for spatial branching processes. In *Seminar on Stochastic Processes, 1987*, eds: E. Çinlar, K.L. Chung, R.K. Getoor. Progress in Probability and Statistics, **15**, 223–241. Birkhäuser, Boston.

[23] Pakes, A.G. (1992). On characterizations via mixed sums. *Austral. J. Statist.* **34**, 323–339.

[24] Seneta, E. (1968). On recent theorems concerning the supercritical Galton-Watson process. *Ann. Math. Statist.* **39**, 2098–2102.

[25] Seneta, E. (1969). Functional equations and the Galton-Watson process. *Adv. Appl. Probab.* **1**, 1–42.

[26] Schuh, H. J. (1982). Seneta constants for the supercritical Bellman-Harris process. *Adv. Appl. Probab.* **14**, 732–751.

[27] Uchiyama, K. (1978). The behaviour of solutions of some non-linear diffusion equations for large time. *J. Math. Kyoto Univ.* **18**, 453–508.

J.D. Biggins, A.E. Kyprianou
Probability and Statistics Section
School of Mathematics and Statistics
The Hicks Building
The University of Sheffield
Sheffield, S3 7RH, U.K.

Progress in Probability, Vol. 40
© 1996 Birkhäuser Verlag Basel/Switzerland

The Growth of an Entire Characteristic Function and the Tail Probabilities of the Limit of a Tree Martingale

QUANSHENG LIU

Mathematics Subject Classification: 60E10, 30D15; 60J80, 60G42, 60K35

Keywords: Branching processes, trees, branching random walks, functional equations; multiplicative chaos, Mandelbrot's martingales; entire characteristic functions, orders, types; tail probabilities; Tauberian theorems

Abstract. For a real random variable whose tail probabilities decrease exponentially, we find the relations among the parameters which describe the growth order of its entire characteristic function, that of its moments, and the decay rate of the tail probabilities. We then evaluate the parameters for a tree martingale which arises in general branching processes, in multiplicative chaos and in infinite particle systems. In particular, we give new results about the asymptotic behaviour of the tail probabilities for age dependent branching processes, for Mandelbrot's martingales and for branching random walks.

Résumé. Pour la limite d'une martingale indexée par un arbre, nous déterminons l'ordre de croissance de sa fonction caractéristique entière, celle de ses moments, et la vitesse de décroissance des probabilités de queue. Cette martingale s'introduit dans les processus de branchement, dans les chaos multiplicatifs ou dans les systèmes infinies de particules. En particulier, nous obtenons de nouveaux résultats concernant l'estimation des probabilités de queue pour un processus de branchement dépendant de l'âge, une martingale de Mandelbrot, ou une marche aléatoire de branchement. Ces résultats sont basés sur certains théorèmes tauberiens de type exponentiel.

1 The growth order of entire characteristic functions and the tail probabilities of random variables

Let $W \not\equiv 0$ be a real random variable (r.v.) and

$$f(z) = E e^{izW} \quad (z \in \mathbf{C}) \tag{1.1}$$

be its characteristic function (c.f.). We say that W has an analytic c.f. if its c.f. coincides with an analytic function at a neighbourhood of 0. It is called to have an entire c.f. if the c.f. coincides with an entire function (see Lukacs (1964)). The following lemma characterizes the class of random variables with an analytic or entire c.f. in terms of the moment generating function, of the tail probabilities or of the moments. As usual, we write

$$f(x) = O(g(x)) \ (x \to \infty) \text{ if } |f(x)| \le Cg(x)$$

for some constant $C > 0$ and all $x > 0$ sufficiently large, and

$$f(x) = o(g(x)) \ (x \to \infty) \text{ if } \lim_{x \to \infty} f(x)/g(x) = 0;$$

for $1 \le p \le \infty$, we denote by $\|\cdot\|_p$ the norm of L^p.

Lemma 1.1. *W has an analytic (resp. entire) c.f. if and only if one of the following equivalent conditions holds :*
  (i) for some (resp. all) $r > 0$, $Ee^{r|W|} < \infty$;
  (ii) for some (resp. all) $r > 0$, $P(|W| > x) = O(e^{-rx})$ as $x \to \infty$;
  (iii) for some (resp. all) $C > 0$ and all $n \geq 1$, $\|W\|_n \equiv [E(|W|^n)]^{1/n} \leq Cn$.

  Of course, the conditions (i), (ii) and (iii) read also respectively
  (i') $\sup\{r \geq 0 : Ee^{r|W|} < \infty\} > 0$ *(resp. $= \infty$),*
  (ii') $R_1 := \liminf_n \frac{\log P(|W| > x)^{-1}}{x} > 0$ *(resp. $= \infty$),*   *and*
  (iii') $\nu_1 := \limsup_n \frac{\|W\|_n}{n} < \infty$ *(resp.$= 0$)*

  Moreover we can prove that [see Proposition 3.4(b) with q = 1]

$$R_1 = \sup\{r \geq 0 : Ee^{r|W|} < \infty\} \text{ and } R_1\nu_1 = 1, \tag{1.2}$$

where and throughout the paper, we make the convention that

$$0\infty = \infty 0 = 1, \ 0^{-1} = \infty, \ \infty^{-1} = 0, \ \frac{\infty}{\infty} = 1, \infty^a = \infty \text{ if } a \in (0, \infty]. \tag{1.3}$$

  The *proof of Lemma 1* is then easy: if the c.f. f coincides with an analytic function in a neighbourhood of 0, then for all small $r > 0$, $f(\pm ir) < \infty$, $Ee^{r|W|} \leq f(ir) + f(-ir) < \infty$ and so *(i)* holds. The equivalence of *(i')*, *(ii')* and *(iii')* is given by (1.2), so we have the equivalence of *(i)*, *(ii)* and *(iii)*. If *(iii)* holds, then the series $\sum i^n E(W^n)z^n/n!$ has positive radius of convergence; this series coincides with the c.f. of W for all $z \in \mathbf{R} = (-\infty, \infty)$ sufficiently near to 0. Therefore the c.f. of W is analytic in a neighbourhood of 0. The proof for the case of entire c.f. is similar.

  As immediate consequences of Lemma 1, we see that W has an analytic but not entire c.f. if and only if $0 < R_1 < \infty$ (which occurs if and only if $0 < \nu_1 < \infty$); and W has an analytic (or entire) c.f. if and only if so does $|W|$.

  Suppose that f is analytic or entire and put

$$M(r) = \max\{|f(z)|; \ |z| = r, z \in \mathbf{C}\}.$$

Then $M(r) \leq \infty$ is also the maximal value of $|f(z)|$ in the disk $|z| \leq r$; $M(r) < \infty$ for all $r > 0$ if f is entire, and $M(r) < \infty$ for some but not all $r > 0$ if f is analytic but not entire.

Definition 1.1. *For a random variable $W \not\equiv 0$ with analytic or entire c.f., let*

$$\rho(resp.\underline{\rho}) = \limsup_{r \to \infty} (resp. \liminf) \frac{\log \log M(r)}{\log r},$$

$$\tau(resp.\underline{\tau}) = \limsup_{r \to \infty} (resp. \liminf) \frac{\log M(r)}{r^\rho},$$

$$Q(resp.\bar{Q}) = \liminf_{x \to \infty} (resp. \limsup) \frac{\log \log P(|W| > x)^{-1}}{\log x},$$

$$R(resp.\bar{R}) = \liminf_{r\to\infty} \;(resp.\limsup) \;\frac{\log P(|W| > x)^{-1}}{x^Q},$$

$$\mu(resp.\underline{\mu}) = \limsup_{n\to\infty} \;(resp.\liminf) \;\frac{\log \|W\|_n}{\log n},$$

$$\nu(resp.\underline{\nu}) = \limsup_{n\to\infty} \;(resp.\liminf) \;\frac{\|W\|_n}{n^\mu},$$

where and throughout the paper we make the convention that whenever we write τ or $\bar{\tau}$ (resp. R or $\bar{R}$, ν or $\bar{\nu}$), we suppose implicitly that ρ (resp. Q, or μ) is finite.

We note that the pairs $(Q, \bar{Q})$, $(R, \bar{R})$, $(\mu, \underline{\mu})$ and $(\nu, \underline{\nu})$ are also well defined for all random variables W whose c.f. are not necessarily analytic. All the numbers defined above lie in $[0, \infty]$. If f is entire, by standard language in the theory of entire functions, ρ (resp. $\underline{\rho}$) is called the *order* (resp. *lower order*) of f; if $\rho < \infty$, then τ (resp. $\underline{\tau}$) is called the *type* (resp. *lower type*) of f with respect to the order ρ; f is called to be of *minimal type, mean type* or *maximal type* according as $\tau = 0$, $0 < \tau < \infty$ or $\tau = \infty$ respectively. If $\rho = \underline{\rho}$, f is called to be of *regular growth* of order ρ; and if $\tau = \underline{\tau}$, f is called to be of *perfectly regular growth* of order ρ and type τ. If f is analytic but not entire, then $\rho = \underline{\rho} = \infty$. Equivalent definitions of the pairs (ρ, τ), (Q, R) and (μ, ν) can be given in the following:

Definition 1.1'. *For all random variables W, we write*

$$\rho = \inf\{a > 0 : M(r) = O(\exp(r^a))\},$$
$$\tau = \inf\{a > 0 : M(r) = O(\exp(ar^\rho))\},$$
$$Q = \sup\{a > 0 : P(|W| > x) = O(\exp(-x^a))\},$$
$$R = \sup\{a > 0 : P(|W| > x) = O(\exp(-ax^Q))\},$$
$$\mu = \inf\{a > 0 : \|W\|_n = O(n^a)\},$$
$$\nu = \inf\{a > 0 : \|W\|_n \le an^\mu \text{ for all n large enough}\}.$$

Since $\|W\|_r$ increases with r, it is easily seen that in the definitions of $(\mu, \underline{\mu})$ and $(\nu, \underline{\nu})$, we can replace the integer n by a *real r* .

Lemma 1.2. *For all (real) random variable W and all $r > 0$,*

$$\frac{1}{2}Ee^{r|W|} \le M(r) \le Ee^{r|W|} \text{ and } M(r) = Ee^{rW} \text{ if } W \ge 0 \text{ a.s.}$$

The proof is easy since $M(r) \ge f(\pm ir)$, $f(ir) + f(-ir) \ge Ee^{r|W|}$ and $|e^z| \le e^{|z|}$ $(z \in \mathbf{C})$.

Due to Lemma 1.2, we can replace $M(r)$ in the definitions of $\rho, \underline{\rho}, \tau$ or $\underline{\tau}$, by the moment generating function $Ee^{r|W|}$ of $|W|$, and the study of the growth of *all* real random variables can be reduced to that of *non-negative* random variables. For example, the entire c.f. of W and $|W|$ have the same order, lower order, type and lower type.

Theorem 1.1. (A characterization of Q and R) *Let W be an arbitrary random variable and let Q and R be defined as in Definition 1.1. Then*

(a) $$Q = \sup\{q \geq 0 : Ee^{|W|^q} < \infty\} \quad and$$

(b) $$R = \sup\{r \geq 0 : Ee^{r|W|^Q} < \infty\} \quad if \quad Q < \infty.$$

This is shown in section 3. For other characterizations of Q, it is convenient to define, for all $q \in [0, \infty)$, that

$$R_q(resp.\bar{R}_q) := \liminf_{x \to \infty} (resp. \limsup) \frac{\log P(|W| > x)^{-1}}{x^q}, \tag{1.4a}$$

and

$$\nu_q(resp.\underline{\nu}_q) := \limsup_n (resp. \liminf) \frac{\|W\|_n}{n^q}. \tag{1.4b}$$

Then for all $0 < q < \infty$, we can prove that (see propositions 3.3 and 3.4) R_q is the radius of the moment generating function of $|W|^q$:

$$R_q = \sup\{r \geq 0 : Ee^{r|W|^q} < \infty\}, \tag{1.4c}$$

and that

$$(qR_q)(\nu_{1/q})^q e = 1 \text{ and } (q\bar{R}_q)(\underline{\nu}_{1/q})^q e \geq 1. \tag{1.4d}$$

Moreover, by Definition 1.1', we have

$$Q = \sup\{q \in [0, \infty) : R_q > 0\} = \sup\{q \in [0, \infty) : R_q = \infty\}$$
$$= \inf\{q \in [0, \infty) : R_q = 0\} = \inf\{q \in [0, \infty) : R_q < \infty\}, \tag{1.4e}$$

$\mu = \inf\{q \in [0, \infty) : \nu_q < \infty\}$, $R = R_Q$, $\nu = \nu_\mu$, etc.

For extreme cases in the following theorem, we recall the convention (1.3).

Theorem 1.2. [Relations among (Q, R), (μ, ν) and (ρ, τ), etc.]
(a) For an arbitrary real r.v. W, we have,

(a1) $$Q\mu = 1 \quad and \quad \bar{Q}\underline{\mu} = 1,$$

where for the last identity we suppose additionally that $Q > 0$ (or $\mu < \infty$);

(a2) $$(QR)(e\nu^Q) = 1 \text{ and } (Q\bar{R})(e\underline{\nu}^Q) \geq 1 \text{ if } 0 < Q < \infty \text{ (or } 0 < \mu < \infty).$$

(b) If W has an analytic or entire c.f., then

(b0) $$\bar{Q} \geq Q \geq 1, \quad \underline{\mu} \leq \mu \leq 1, \quad \rho \geq \underline{\rho} \geq 1;$$

$(b1)$
$$\frac{1}{Q} = \mu = 1 - \frac{1}{\rho} \geq 1 - \frac{1}{\underline{\rho}} \geq \underline{\mu} = \frac{1}{\underline{Q}} \geq \frac{\rho}{Q} - \frac{1}{\rho},$$

where for the last inequality we suppose additionally that $\rho < \infty$;

(b2) If $1 < Q < \infty$ (or $0 < \mu < 1$ or $1 < \rho < \infty$), then

$$(QR)^{-1/Q} = e^\mu \nu = (\rho\tau)^{1/\rho} \geq (\underline{\rho}\underline{\tau})^{1/\rho} \geq e^\mu \underline{\nu} \geq (\underline{Q}\bar{R})^{-1/Q};$$

(b3) If $1 < Q < \infty$ (or $1 < \rho < \infty$ or $0 < \mu < 1$) and $0 < \underline{\tau} \leq \tau < \infty$, then

$$\frac{\lambda}{\bar{\lambda}} \leq \left(\frac{R}{\bar{R}}\right)^{1/Q} \leq \frac{\nu}{\bar{\nu}} \leq \left(\frac{\underline{\tau}}{\tau}\right)^{1/\rho},$$

where $0 < \lambda \leq \rho\underline{\tau} \leq \rho\tau \leq \bar{\lambda} < \infty$ are the two positives solutions of the equation

$$x - Rx^Q = \underline{\tau} , \tag{1.5a}$$

and the two solutions are equal if and only if $\underline{\tau} = \tau$.

The conclusion $\rho \geq 1$ in (b0) is a well-known result of Paul Lévy.

The most interesting aspect of the theorem is perhaps the symmetry of the results. Let us make some comments on the conclusion that

$$\frac{1}{\rho} + \frac{1}{Q} = 1 \tag{1.6a}$$

and

$$(\rho\tau)^{1/\rho}(QR)^{1/Q} = 1 \tag{1.6b}$$

(these are contained in (b1) and (b2) above). It gives a criterion for a distribution on $\mathbf{R}(= (-\infty, \infty))$ to have an entire c.f. of given order, or of given order and given type if the order is finite; it unifies all the Theorems 6.1, 7.1-7.3, 8.1, 9.1, and 9.2.2 of Ramachandran (1962). Before Ramachandran, less precise results were given by E. Lukacs (1964, pp. 151-155) and D. Dugué (1939). Defining Q and R as in (a) and (b) of Theorem 1.1, Harris (1948) proved the identity (1.6a), conjectured (1.6b) and proved it in the case where ρ is rational. Of course, our conclusion here shows that *Harris' conjecture is right.*

Sometimes it is convenient to write (1.6a) and (1.6b) as

$$\rho = Q/(Q-1) \text{ and } \tau = (1 - Q^{-1})(QR)^{-1/(Q-1)}; \tag{1.6c}$$

therefore, by the symmetry in (1.6a) and (1.6b), exchanging the pairs (Q,R) and (ρ, τ) in (1.6c) we obtain

$$Q = \rho/(\rho-1) \text{ and } R = (1 - \rho^{-1}))(\rho\tau)^{-1/(\rho-1)}. \tag{1.6d}$$

The corresponding conclusion for $\underline{\tau}$ and $\overline{R}$ is

$$\underline{\tau} \geq (1 - Q^{-1})(Q\overline{R})^{-1/(Q-1)} \quad \text{and} \quad \overline{R} \geq (1 - \rho^{-1}))(\rho\underline{\tau})^{-1/(\rho-1)}. \qquad (1.6e)$$

As a consequence of Theorem 1.2, we obtain

Theorem 1.3. (Preservation of regular growth)
(a) If W has an entire c.f. then

$(a1) \qquad 1 < Q \leq \bar{Q} < \infty \iff 0 < \underline{\mu} \leq \mu < 1 \iff 1 < \underline{\rho} \leq \rho < \infty;$

$(a2) \qquad 1 < Q < \infty \iff 0 < \mu < 1 \iff 1 < \rho < \infty;$

$\qquad\qquad\qquad Q = 1 \iff \mu = 1 \iff \rho = \infty;$

$\qquad\qquad\qquad Q = \infty \iff \mu = 0 \iff \rho = 1;$

$(a3) \qquad\qquad\qquad Q = \bar{Q} \iff \mu = \underline{\mu} \iff \rho = \underline{\rho}\,.$

In (a3), for the implications "$\Leftarrow$", we suppose moreover that $\rho < \infty$.
(b) If additionally $1 < Q < \infty$ (or $0 < \mu < 1$ or $1 < \rho < \infty$), then

$(b1) \qquad 0 < R \leq \bar{R} < \infty \iff 0 < \underline{\nu} \leq \nu < \infty \iff 0 < \underline{\tau} \leq \tau < \infty;$

$(b2) \qquad 0 < R < \infty \iff 0 < \nu < \infty \iff 0 < \tau < \infty;$

$\qquad\qquad\qquad R = 0 \iff \nu = \infty \iff \tau = \infty;$

$\qquad\qquad\qquad R = \infty \iff \nu = 0 \iff \tau = 0;$

$(b3) \qquad\qquad\qquad R = \bar{R} \iff \nu = \underline{\nu} \iff \tau = \underline{\tau}.$

In (b3), for the implications "$\Leftarrow$" we suppose moreover that $0 < \underline{\tau} \leq \tau < \infty$.

(a3) and (b3) give criterions for a distribution on $\mathbf{R}$ to have an entire c.f. of regular growth or perfectly regular growth.

Let us give some more comments on the equivalence between $R = \bar{R}$ and $\tau = \underline{\tau}$. This is typically a Tauberian theorem of exponential type. For a positive r.v. $W \geq 0$, it says that

$$\lim_{r \to \infty} \frac{1}{r^p} \log Ee^{rW} = t \iff \lim_{r \to \infty} \frac{1}{r^{p^*}} \log P(W \geq r)^{-1} = t^*, \qquad (1.7a)$$

where $p > 1, p^* > 1, t > 0, t^* > 0$ are any real numbers related by

$$\frac{1}{p} + \frac{1}{p^*} = 1 \quad \text{and} \quad (pt)^{1/p}(p^*t^*)^{1/p^*} = 1. \qquad (1.7b)$$

(1.7) contains Davies' (1976) theorem, and can be regarded as a special case of Kasahara's (1987) Tauberian theorem.

The implication "$\Rightarrow$" in (1.7) is also a large deviation result. The relation between the asymptotical behaviour of exponential moments and probability tails (of a family of r.v.) is the core of "Large Deviations Techniques". The two functions ruling these asymptotics are then Legendre-Fenchel-Young conjugate (Dembo-Zeitouni). In our situation, we remark that the two parts in (1.7) can be re-written respectively as

$$\lim_{r\to\infty} \frac{1}{ptr^p} \log Ee^{rxW} = \frac{x^p}{p}, \qquad\qquad \forall x > 0,$$

and

$$\lim_{r\to\infty} \frac{1}{ptr^p} \log P\Big(\frac{rW}{ptr^p} > y\Big)^{-1} = \frac{y^{p*}}{p*}, \qquad\qquad \forall y > 0.$$

Therefore, because the functions $x \mapsto \frac{x^p}{p}$ and $y \mapsto \frac{y^{p*}}{p*}$ are Legendre conjugate, the implication "$\Rightarrow$" in (1.7) can also be obtained from the Gartner-Ellis theorem (Dembo-Zeitouni (1993), th 2.3.6), and the converse "$\Leftarrow$" from the usual trick of Varadhan's lemma.

Some results in Theorems 2(b) and 3(b) can also be deduced from Kasahara's (1987) tauberian theorem (by using lemma 1.2 or corollary 1.1 in the preceding); but I prefer to give direct proofs, not only for reader' convenience, or to make the study more systematic and the paper more self-contained, but also because the proofs given here are surprisingly simple.

2 Evaluations for tree martingales; examples and applications

2.1. A functional equation; evaluations for tree martingales.

Let $N \geq 0$ be a random integer and $\{A_i\}$ be a sequence of non-negative random variables. The joint distribution of $\{N; A_1, A_2, \ldots\}$ is arbitrary. We consider the functional equation

$$Z = \sum_{i=1}^{N} A_i Z_i, \qquad\qquad (2.1a)$$

where the empty sum is taken to be 0 (which happens when $N = 0$), $Z_i \geq 0$ $(i = 1, 2, \ldots)$ are independent random variables distributed as $Z \geq 0$, and also independent of N and of $\{A_i\}$; the distribution of Z is unknown. If Φ is the Laplace transform of Z, the equation is equivalent to

$$\Phi(t) = E \prod_{i=1}^{N} \Phi(tA_i), \quad t \geq 0, \qquad\qquad (2.1b)$$

where (and throughout the paper) the product is taken over all the indices i such that $A_i > 0$, and the empty product is taken to be 1.

Kahane and Peyrière (1976), and Guivarc'h (1990) studied the fixed points of the transformation T in the case where N is constant and the A_i ($1 \leq i \leq N$) are i.i.d. Their works were motivated by questions raised by Mandelbrot related to a model of turbulence of Yaglom. Holley and Liggett (1981) studied the same problem in the case where N is constant and the A_i are fixed multiples of one random variable, and Durrett and Liggett (1983) considered the more general case where N is constant but the A_i have arbitrary joint distribution. Their works were motivated by a number of problems in infinite particle systems. Closely related results are given in Kahane (1987), Ben Nasr (1987), Holley and Waymire (1992), Collet and Koukiou (1992), Franchi (1993) and Chauvin and Rouault (1996), etc.

If $1 < m = EN < \infty$ and $A_i = 1/m$ ($1 \leq i \leq N$), then the equation (2.1b) reduces to the Poincaré functional equation $\Phi(u) = E\Phi^N(u/m)$, which arises in the Galton-Watson process. Similar equations (which are always special cases of our equation $\Phi = T\Phi$) arise in age-dependent branching processes or branching random walks. The study of these equations has been important, since it gives the limit behaviour of the population sizes of the associated processes. Many authors have contributed to it, see for example Harris (1948), Kesten and Stigum (1966), Seneta (1968 and 1969), Athreya (1971), Doney (1972 and 1973), Bingham and Doney (1974 and 1975) and Biggins (1977).

The equation (2.1), in its various forms, was also used to study some fractal sets or flows in networks, implicitly or directly by Mauldin and Williams (1986), Falconer (1986 and 1987) and Liu (1993 and 1996).

Let

$$\tilde{N} := \sum_{i=1}^{N} 1_{\{A_i > 0\}}$$

be the number of non-zero terms of A_i, $1 \leq i \leq N$. To simplify the discussion, we suppose throughout the paper that

$$P(\tilde{N} = 0 \text{ or } 1) < 1, \quad P(\forall i \in \{1, \ldots, N\}, \ A_i = 0 \text{ or } 1) < 1, \tag{2.2}$$

$$E\tilde{N} < \infty \text{ and } E\sum_{i=1}^{N} A_i \log^+ A_i < \infty, \tag{2.3}$$

where $\log^+ x = \max(0, \log x)$. For $x \in [0, \infty)$, write

$$S(x) := \sum_{i=1}^{N} A_i^x \text{ and } \rho(x) := ES(x),$$

where (and throughout) the sum is taken over all the i's such that $A_i > 0$. The function ρ is defined on $[0, \infty)$ with values in $[0, \infty]$. Then

$$\rho(x) < \infty \text{ and } \rho'(x) = E\sum_{i=1}^{N} A_i^x \log A_i < \infty$$

exists for all $x \in (0, 1]$ (and $\rho'(1)$ denotes the left derivative); and ρ is strictly convex on $(0, 1)$.

Write $\mathbf{N} = \{0, 1, 2, \ldots\}$, $\mathbf{N}_+ = \{1, 2, \ldots\}$, and, for all sequences $\sigma = \sigma_1\sigma_2 \cdots \sigma_N \in \mathbf{N}_+^n$, $n \geq 0$, $\mathbf{N}_+^0 = \{\emptyset\}$, take

$$(N_\sigma; A_{\sigma 1}, A_{\sigma 2}, \cdots, A_{\sigma N_\sigma})$$

as independent copies of $(N; A_1, \ldots, A_n)$. For $n > 0$, put

$$W_n := \sum_{\sigma_1, \cdots, \sigma_n} A_{\sigma_1} \ldots A_{\sigma_1 \cdots \sigma_n},$$

where the sum is taken over all the nodes $\sigma = \sigma_1 \cdots \sigma_n$ at the level n of the associated Galton-Watson tree, and let $\mathcal{F}_n$ be the σ-field generated by

$$\{(N_\sigma; A_{\sigma 1}, A_{\sigma 2}, \ldots, A_{\sigma N_\sigma}) : \ |\sigma| \leq n - 1\},$$

where $|\sigma|$ denotes the length of σ; the length of $\emptyset$ is taken to be 0. Then $\{(W_n, \mathcal{F}_n) : n \geq 1\}$ is a martingale. We call it a *tree martingale*, following Falconer (1987). Tree martingales are closely related to branching random walks. Put

$$W = \lim_{n \to \infty} W_n, \tag{2.4}$$

which exists and is finite a.s. by the martingale convergence theorem, and $EW \leq 1$ by Fatou's lemma. It is easily seen that W is a solution of (2.1a).

The following results are known.

Theorem 2.0. (Liu 1995) *Assume (2.2) and (2.3). Then the equation (2.1a) has a non-trivial solution if and only if for some $\alpha \in (0, 1]$, $\rho(\alpha) = 1$ and $\rho'(\alpha) \leq 0$; it has a nontrivial solution with finite mean if and only if*

$$ES(1)\log^+ S(1) < \infty, \ \rho(1) = 1 \ and \ \rho'(1) < 0. \tag{2.5}$$

There is at most one nontrivial solution $Z \geq 0$ which satisfies $EZ = 1$; if (2.4) holds, then the random variable W defined in (2.4) is the unique such solution; moreover, for all $p = 2, 3 \ldots$ $EW^p < \infty$ if and only if $\rho(p) < 1$ and $E[(S(1)^p] < \infty$.

Here and throughout the paper, when we speak of the uniqueness of solution of (2.1a), we identify of course random variables which have the same distribution. In this paper, we only consider solutions of finite moments of all orders; by Theorem 2.0, such solutions are parametrized by their means. We remark that if $\rho(p) < 1$ for all $p = 2, 3, \cdots$, then $\| \sup_{1 \leq i \leq N} A_i \|_p \leq 1$, and so $\| \sup_{1 \leq i \leq N} A_i \|_\infty \leq 1$. Therefore, again by Theorem 2.0, W *has moments of all orders if and only if*

$$P(\forall i \in \{1, \ldots, N\}, \ A_i \leq 1) = 1 \ and \ E[S(1)^p] < \infty \ for \ all \ p > 1. \tag{2.6}$$

Assume (2.6) and put

$$\beta = \inf\{b \in [0,1) : S\big((1-b)^{-1}\big) \leq 1 \text{ a.s.}\}, \quad \text{where } \inf \emptyset := 1. \tag{2.7a}$$

Then $\beta \in [0,1]$; $\beta = 0$ if and only if $P[S(1) = 1)] = 1$; and β is the least solution in $[0,1)$ of the equation

$$\|S(\frac{1}{1-\beta})\|_\infty = 1 \tag{2.7b}$$

if $\beta < 1$ and if there is (at least) a solution in $[0,1)$. Moreover, if $x_0 > 0$ and if $x \geq x_0$ and $y \geq x_0$, then

$$\big| \, \|S(x)\|_\infty - \|S(y)\|_\infty \, \big| \leq \|S(x) - S(y)\|_\infty \leq |x - y| \, \| \sum_{i=1}^{N} A_i^{x_0} \, |\log A_i| \, \|_\infty.$$

Therefore, if $\| \sum_{i=1}^{N} A_i^{x_0} \, |\log A_i| \, \|_\infty < \infty$ for some $x_0 > 0$, then the function $x \to \|S(x)\|_\infty$ is Lipschitzian on $[x_0, \infty)$. In particular, if $\|N\|_\infty < \infty$, then $\|S(x)\|_\infty$ is continuous on $(0, \infty)$, and so β satisfies (2.7b) if $\beta < 1$.

     The following results will be shown in section 4.

Theorem 2.1. (Tail behaviour of W) *Assume (2.2), (2.3), (2.5), (2.6) and $P[S(1) = 1)] < 1$. Let W and β be defined as in (2.4) and (2.7). Then W is the unique solution $Z \geq 0$ of (2.1a) with $EZ = 1$, and, defining $(Q, \bar{Q})$, $(R, \bar{R}) \dots$ as in Definition 1.1, we have*

(a) If $\|N\|_\infty < \infty$, then

$$Q = \bar{Q} = 1/\beta \quad \text{and} \quad R > 0, \tag{2.8a}$$

$$\mu = \underline{\mu} = \beta \quad \text{and} \quad \nu < \infty, \tag{2.8b}$$

$$\rho = \underline{\rho} = 1/(1-\beta) \text{ and } \tau < \infty; \tag{2.8c}$$

in other words, for all $\varepsilon > 0$, there are some constants $c_i \in (0, \infty)$, $1 \leq i \leq 6$, such that for all $r > 0$ sufficiently large, we have

$$\exp\{-c_1 r^{\beta^{-1}+\varepsilon}\} \leq P(W > r) \leq \exp\{-c_2 r^{\beta^{-1}}\}, \tag{2.8d}$$

$$c_3 r^{\beta-\varepsilon} \leq \|W\|_r \leq c_4 r^\beta, \tag{2.8e}$$

$$c_5 r^{(1-\beta)^{-1}-\varepsilon} \leq \log E e^{rW} \leq c_6 r^{(1-\beta)^{-1}}. \tag{2.8f}$$

If additionally one of the following conditions holds:
*   (i) there exists an integer $n > 1$ such that*

$$\beta = \log \|S(1)1_{N=n}\|_\infty / \log n \tag{2.9a}$$

and that for some constants $a \geq 0$, $c > 0$ and for all $x > 0$ sufficiently small,

$$P\{\frac{S(1)1_{\{N=n\}}}{\|S(1)1_{\{N=n\}}\|_\infty} > 1 - x\} \geq cx^a; \tag{2.9b}$$

(ii) $\beta < 1$ and there exist some constants $0 < \delta < 1$, $a \geq 0$ and $c > 0$ such that, for all $x > 0$ sufficiently small,

$$P\{S(\frac{1}{1-\beta}) > 1 - x \text{ and } A_i \leq \delta \text{ for all } 1 \leq i \leq N\} \geq cx^a, \tag{2.9c}$$

then (2.8) can be improved as

$$Q = \bar{Q} = 1/\beta \quad \text{and} \quad 0 < R \leq \bar{R} < \infty, \tag{2.10a}$$

$$\underline{\mu} = \mu = \beta \quad \text{and} \quad 0 < \underline{\nu} \leq \nu < \infty, \tag{2.10b}$$

$$\rho = \underline{\rho} = 1/(1-\beta) \quad \text{and} \quad 0 < \underline{\tau} \leq \tau < \infty; \tag{2.10c}$$

namely, for some constants $c_i \in (0, \infty)$ $(1 \leq i \leq 6)$ and for all $r > 0$ sufficiently large,

$$\exp\{-c_1 r^{\beta^{-1}}\} \leq P(W > r) \leq \exp\{-c_2 r^{\beta^{-1}}\}, \tag{2.10d}$$

$$c_3 r^\beta \leq \|W\|_r \leq c_4 r^\beta, \tag{2.10e}$$

$$c_5 r^{(1-\beta)^{-1}} \leq \log Ee^{rW} \leq c_6 r^{(1-\beta)^{-1}}. \tag{2.10f}$$

(b) If $\|N\|_\infty = \infty$, then

$$Q \leq 1/\beta, \ \mu \geq \underline{\mu} \geq \beta \text{ and } \rho \geq 1/(1-\beta).$$

Here we make the convention that when we write (2.8f) or (2.10f), we suppose implicitly that $\beta < 1$.

(2.10a) gives the asymptotic behavior of the tail probabilities of W; (2.10b) concerns the growth rate of the moments of W; and (2.10c) implies that W has an entire c.f. of regular growth with order $\rho = 1/(1-\beta)$ and type $\tau \in (0, \infty)$.

Remark 2.1. (1) The condition (2.9a) holds usually (cf. Liu 1993, and also section 2.2 below); (2.9b) holds evidently for $a = 0$ if

$$P\{S(1) = \|S(1)1_{\{N=n\}}\|_\infty \text{ and } N = n\} > 0$$

or, more particularly, if $P(N = n) > 0$ and if A_i $(1 \leq i \leq N)$ take only a finite number of values a.s. on the event $\{N = n\}$. Moreover, in Theorem 2.1, the conditions (2.9b) and (2.9c) can be weakened respectively to

$$\prod_{i=1}^\infty \frac{\|S(1)1_{\{N=n\}}\|_{n^i}}{\|S(1)1_{\{N=n\}}\|_\infty} > 0, \tag{2.9b'}$$

and

$$\prod_{k=1}^\infty \|S(1/(1-\beta))1_{\{\sup_{1\leq i \leq N} A_i \leq \delta\}}\|_{p^k} > 0 \tag{2.9c'}$$

for some $0 < \delta < 1$ and $1 < p < \delta^{-1/(1-\beta)}$.

(2) The condition (ii) of Theorem 2.1 holds if one of the following conditions is satisfied:

(a) $\beta > 0$ and, for all i, A_i takes only finitely many values a.s.,

(b) $P(S(\frac{1}{1-\beta}) = 1$ and $\tilde{N} > 1) > 0$,

(c) $P(S(\frac{1}{1-\beta}) = 1$ and $\sup_{1 \leq i \leq N} A_i < 1) > 0$,

(d) $N = n \geq 2$ is a.s. a constant and $(A_1^{1/(1-\beta}, \ldots, A_n^{1/(1-\beta}$ has a density $d(x_1, \ldots, x_n)$ with respect to the Lebesgue measure on $\Delta := \{(x_1, \ldots, x_n) \in [0, 1]^n : x_1 + \ldots + x_n \leq 1\}$ which satisfies the following property: there exists a point $(x_1^0, \ldots, x_n^0) \in \Delta$ with $x_i^0 > 0 (1 \leq i \leq n)$ and $x_1^0 + \ldots + x_n^0 = 1$ such that for some constants $c > 0$, $a \geq 0$ and for all $0 \leq x_i < x_i^0 (1 \leq i \leq n)$ sufficiently near to x_i^o,

$$d(x_1, \ldots, x_n) \geq c[1 - (x_1 + \ldots + x_n)]^a.$$

(3) In the study of fractal sets, to calculate the exact Hausdorff dimension functions, we often need sufficient conditions to ensure that $R < \infty$, see Graf, Mauldin and Williams (1988) and Liu (1993); of course, both (i) and (ii) of Theorem 2.1 are such conditions; and each of them covers all the examples of Graf et al. (except their counter-examples) and of Falconer (1986). Graf et al. (1988) use the condition (d) above in the case where $a = 0$.

(4) The condition (i) or (ii) in the theorem cannot be removed; moreover, we even cannot remove the corner constraint "$A_i \leq \delta$ for all $1 \leq i \leq N$" in (2.9c). As counter-examples, for the first, we can take Example 6.10 of Graf et al. (1988); and for the second, we take their Example 6.11.

The case where $\{A_i\}$ are independent and identically distributed (i.i.d.) is of particular interest.

Corollary 2.1. *Assume (2.2), (2.3), (2.5), (2.6) and $P[S(1) = 1)] < 1$, and let W be defined as in (2.4). If $\|N\|_\infty < \infty$ and if $\{A_i\}$ are i.i.d.r.v.'s which are also independent of N, then (2.8a)-(2.8f) hold with*

$$\beta = 1 + \frac{\log \|A_1\|_\infty}{\log \|N\|_\infty}; \tag{2.11a}$$

if additionally for some constants $c > 0$, $a \geq 0$ and all $x > 0$ sufficiently small,

$$P(\frac{A_1}{\|A_1\|_\infty} > 1 - x) \geq cx^a, \tag{2.11b}$$

then (2.10a)-(2.10f) hold.

2.2. Examples and applications.

We now give some examples for applications of Theorems 1.2 and 2.1.

Example 2.1. (The Galton-Watson process) Let $N \geq 0$ be an integer-valued random variable with $1 < m := EN < \infty$ and put $A_i = 1/m$ for $1 \leq i \leq N$. Then

$$W = \lim_n Z_n/m^n. \tag{2.12}$$

Z_n being the associated Galton-Watson process with $Z_0 = 1$ and offspring distribution given by N. If $\|N\|_\infty < \infty$ and N is not a.s constant, then

$$\beta = 1 - \log m / \log \|N\|_\infty \in (0, 1). \tag{2.13}$$

The condition (2.11b) holds evidently for $c = 1$ and $a = 0$; so by Corollary 2.1, W satisfies (2.10a)-(2.10f). The conclusion (2.10c), namely that

$$\rho = \underline{\rho} = 1/(1 - \beta) = \log \|N\|_\infty / \log m \text{ and } 0 < \underline{\tau} \leq \tau < \infty,$$

is a famous result of Harris (1948). In fact, Harris proved that

$$\log M(r) \sim r^\rho H(r), \ r \to \infty,$$

where the Harris function $H(.)$ is continuous, strictly positive and multiplicatively periodic with period m. It was in this context that Harris conjectured (1.6.b) to deduce the value of R from the value of τ (Harris defined R as in (b) of Theorem 1.1). By Harris' result,

$$\underline{\tau} = \min_{1 \leq r \leq m} H(r) \text{ and } \tau = \max_{1 \leq r \leq m} H(r).$$

Recently, Biggins and Bingham (1993) proved that

$$\liminf_{x \to \infty} \frac{\log P(|W| > x)^{-1}}{x^{(\rho/\rho-1)}} \geq \frac{(1 - \rho^{-1})}{(\rho\tau)^{1/(\rho-1)}}$$

and

$$\limsup_{x \to \infty} \frac{\log P(|W| > x)^{-1}}{x^{(\rho/\rho-1)}} \leq \frac{(1 - \rho^{-1})}{(\rho\underline{\tau})^{(1/\rho-1)}};$$

but the opposite inequalities also hold by Theorem 1.2(b2) (cf.(1.6c)-(1.6e)), so we have in fact the corresponding *equalities*. The conclusion that $\mu = \underline{\mu} = \beta$ was also proved in Liu (1996) to obtain exact Hausdorff dimension functions for branching sets.

Example 2.2. (The Bellman-Harris process) Let $(Z_t : t \geq 0; Z_0 = 1)$ be an age-dependent branching process characterized by a generating function $f(s) = \sum p_k s^k$ governing particle production and a distribution G on $[0, \infty)$ describing particle life length. The process is initiated at time $t = 0$ with a parent particle which lives for time L_0 and then produces N particles according to $\{p_k\}$. These live for times $L_{11}, \ldots L_{1N}$, and then produce offspring according to $\{p_k\}$, etc. The L's are independent random variables with distribution G; particle production is independent of the present state or past history of the process; and the life time and particle production variables are independent (cf. Athreya and Ney (1972), p. 137). Suppose that $1 < EN < \infty$ (supercritical), then

$$W := \lim_{t \to \infty} \frac{Z_t}{EZ_t} \tag{2.14}$$

exists a.s. (cf. Athreya-Kaplan (1976)) and its Laplace transform satisfies

$$\Phi(s) = \int_0^\infty f(\Phi(se^{-\alpha y}))dG(y), \quad s \geq 0, \tag{2.15a}$$

where α (the parameter of Malthus) is the unique number in $(0, \infty)$ satisfying

$$(EN) \int_0^\infty e^{-\alpha y}dG(y) = 1.$$

(cf. Athreya (1971)). The equation can also be written as

$$\Phi(s) = E\Phi^N(se^{-\alpha L}), \quad s \geq 0, \tag{2.15b}$$

where $E(Ne^{-\alpha L}) = 1$. So this is a special case of (2.1b) with $A_i = e^{-\alpha L}$ for all $i \geq 1$ and $E\sum_{i=1}^N A_i = 1$, $\{A_i : i \geq 1\}$ being independent of N. Athreya (1971) proved that if $EN\log^+ N < \infty$, then W is the unique solution of the equation satisfying $EW = 1$.

Corollary 2.2. *Let (Z_t) be the age-dependent branching process introduced above and let W be defined in (2.14). Put*

$$\beta = 1 - \frac{\log \|e^{-\alpha L}\|_\infty}{\log \|N\|_\infty}. \tag{2.16}$$

If $\|N\|_\infty < \infty$, then (2.8a)-(2.8f) hold; if additionally for some constants $c > 0$, $a \geq 0$ and all $x > 0$ sufficiently small,

$$P(L < l + x) \geq cx^a, \quad where \quad l := ess.inf \ L. \tag{2.17}$$

then (2.10a)-(2.10f) hold.
If $\|N\|_\infty = \infty$, then $Q \leq 1$, $\mu \geq \underline{\mu} \geq 1$ and $\rho = \infty$.

    We remark that the condition (2.17) holds evidently for $a = 0$ if $P(L = l) > 0$.

    The result follows by Theorem 2.1. In fact, it is easily verified that the β defined in (2.7) can be given by (2.16); so (2.9a) holds for $n = \|N\|_\infty$. (2.9b) reduces to (2.17) (the constant c may change).

Example 2.3. (The Mandelbrot's Martingale) Let $N = n \geq 2$ be a constant integer, and let $\{A_i : 1 \leq i \leq n\}$ be a sequence of i.i.d. positive random variables with $nEA_1 = 1$ and $P(nA_1 = 1) < 1$. Then $\{W_n\}$ is the martingale of Mandelbrot for multiplicative chaos.

Corollary 2.3. *Let W be the limit of Mandelbrot's martingale $\{W_n\}$. Then*
(a) W is non-degenerate if and only if $EA_1 \log A_1 < 0$.
(b) If $EA_1 \log A_1 < 0$, then $EW = 1$ and, $EW^k < \infty$ for all $k > 1$ if and only if $\|A_1\|_\infty \leq 1$. In the case where $\|A_1\|_\infty \leq 1$, (2.8a)-(2.8f) hold for

$$\beta = \frac{\log(n\|A_1\|_\infty)}{\log n}; \tag{2.18}$$

if additionally for some constants $c > 0$, $a \geq 0$ and all $x > 0$ sufficiently small,

$$P\left(\frac{A_1}{\|A_1\|_\infty} > 1 - x\right) \geq cx^a \tag{2.19}$$

then (2.10a)-(2.10f) hold.

The result that $\mu = \underline{\mu} = \beta$ in (2.8b) was proved by Kahane and Peyrière (1976 th.3). We remark that in the statement of their theorem, there is an additional condition that

$$P(A_1 = 1) < 1/n;$$

but it is equivalent to $P(A_1 = 0 \text{ or } 1) < 1$, and so is contained in the hypothesis $EA_1 \log A_1 < 0$. In fact, noting that $EA_1 = 1/c$, we have the implications

$$P(A_1 = 1) = \frac{1}{n} \Leftrightarrow P(A_1 = 0 \text{ or } 1) = 1, \tag{2.20a}$$

and, since we have always $P(A_1 = 1) \leq EA_1 = 1/n$, it follows that

$$P(A_1 = 1) < 1/n \Leftrightarrow P(A_1 = 0 \text{ or } 1) < 1. \tag{2.20b}$$

Example 2.4 (The Crump-Mode process) We consider a general branching process $\{Z(t) : t \geq 0\}$ in the sense of Crump and Mode (1968-1969) with a single ancestor $Z(0) = 1$. Each individual reproduces independently; for any given parent individual the instants of birth of offspring are represented by the jumps of a counting process $\{N(t) : t \geq 0\}$ with $N(0) = 0$ and $N(\infty) < \infty$ which increases by one at the instants of birth of offspring; this process and the life time L of the parent may be dependent. We assume throughout that *either* $1 < EN(\infty) < \infty$. Let $\alpha > 0$ be the unique number satisfying

$$E \int_0^\infty e^{-\alpha x} dN(x) = 1. \tag{2.21}$$

It was proved by Doney (1972) that the limit (in distribution)

$$W : \overset{d}{=} \lim_{t \to \infty} \frac{Z(t)}{EZ(t)} \tag{2.22}$$

exists, and satisfies the functional equation

$$\Phi(s) = E \exp\left\{ \int_0^\infty \log \Phi\left(se^{-\alpha x}\right) dN(x) \right\} \ (s \geq 0), \tag{2.23}$$

where $\Phi(s) = Ee^{-sW}$. Writing $N = N(\infty)$ for the total number of offspring of a given parent, $t_1, \ldots t_N$ for the successive instants of their births, and

$$A_i = e^{-\alpha t_i}, \ (1 \leq i \leq N) \tag{2.24}$$

we see that (2.23) can be reformulated as

$$\Phi(s) = E \prod_{i=1}^{N} \Phi(sA_i). \tag{2.25}$$

So Theorem 2.1 also applies, yielding that, for example, if $\|N\|_\infty < \infty$, then W [defined in (2.22)] satisfies (2.8a)-(2.8f).

Example 2.5 (Branching random walks) A branching random walk on $\mathbf{R}$ can be described in the following way. An initial ancestor, who forms the zeroth generation, is created at the origin. His children form the first generation and their positions on $\mathbf{R}$ are described by a point process Z^1 on $\mathbf{R}$. Thus Z^1 is a random locally finite counting measure. The people in the nth generation give birth independently of one another and of the preceding generations to form the $(n+1)$th generation. The point process describing the displacements of the children of a person from that person's position has the same distribution as Z^1. Let $\{z_r^n\}$ be an enumeration of the positions of the people in the nth generation, and Z^n be the point process with atoms $\{z_r^n\}$. Define

$$m(\theta) := E \sum_r \exp(-\theta z_r^1) = E \int \exp(-\theta t)\, dZ^1(t). \tag{2.26}$$

We assume $m(0) > 1$ and $m(\theta) < \infty$ for some fixed θ. The generation size $Z^n(-\infty, \infty)$ form a supercritical Galton-Watson process. It is known (see Biggins 1977) that

$$W^n(\theta) := m(\theta)^{-n} \sum_r exp(-\theta z_r^n) \tag{2.27a}$$

is a martingale with respect to the σ-field $\mathcal{F}_n$ generated by the births in the first n generations, and the Laplace transform $E\exp(-sW(\theta))$ of the limit

$$W(\theta) := \lim_n W^n(\theta) \tag{2.27b}$$

satisfies the functional equation

$$\Phi(s) = E \prod_{i=1}^{N} \Phi(sA_r), \ \ s \geq 0, \tag{2.28}$$

where $A_r = m(\theta)^{-1} \exp(-\theta z_r^1)$ and $E \sum_r A_r = 1$. Therefore Theorem 2.1 applies again in the present setting.

3 Proofs of Theorems 1.1 and 1.2.

The following inequalities will be frequently used.

Lemma 3.1. *(a) For all x and $r > 0$,*

$$P(|W| > x) \leq 2M(r)e^{-rx}.$$

(b) For all $r,x > 0$, if $\eta \equiv \eta(r,x) := e^{rx}/Ee^{r|W|} < 1$, then

$$P(|W| > x) \geq \frac{1}{2}(1 - \eta)^2 M^2(r)/M(2r).$$

Proof. Part (a) follows from Markov's inequality and Lemma 1.2 of section 1. For part (b) we remark that for any r.v. $X \geq 0$ and any real $0 < \lambda < 1$,

$$P(X > \lambda EX) \geq (1 - \lambda)^2 (EX)^2/E(X^2). \tag{3.1}$$

(See Kahane (1985), chap.1, p.8) take $X = e^{r|W|}$ and $\lambda = \eta$, and apply Lemma 1.2 again.

Proposition 3.2. *If W has an entire or analytic c.f., then*

$$\frac{\rho}{\rho - 1} \leq \overline{Q} \leq \frac{\rho}{\underline{\rho} - 1},$$

where for the right inequality we suppose additionally that $\rho < \infty$.

Proof. (a) To prove that $\rho/(\rho - 1) \leq \overline{Q}$, we suppose $\overline{Q} < \infty$. Hence, for all $\epsilon > 0$, there is some $r_\epsilon > 0$ such that for all $r > r_\epsilon$,

$$P(|W| > r) \geq \exp\{-r^{\overline{Q}+\epsilon}\}.$$

For $0 < a < 1/(\overline{Q} - 1 + \epsilon)$, let $x = r^a$. Using Lemma 3.1(a) we get

$$M(r) \geq \frac{1}{2}\exp\{r^{a+1}(1 - r^{(\overline{Q}-1+\epsilon)a-1})\}$$

for $r > 0$ large enough. Thus $\rho \geq a + 1$. Letting consecutively $a \to 1/(\overline{Q} - 1 + \epsilon)$ and $\epsilon \to 0$, we obtain $\rho \geq \overline{Q}/(\overline{Q} - 1)$; therefore $\overline{Q} \geq \rho/(\rho - 1)$.

(b) We then prove that $\overline{Q} \leq \frac{\rho}{\underline{\rho}-1}$ if $1 < \underline{\rho} < \infty$. We want to apply Lemma 3.1(b). For all $0 < \epsilon < \underline{\rho} - 1$, there is some $r_\epsilon > 0$ such that for all $r > r_\epsilon$,

$$\exp\{r^{\underline{\rho}-\epsilon}\} \leq M(r) \leq \exp\{r^{\rho+\epsilon}.\}$$

For all $0 < b < \rho - 1 - \epsilon$ and $x = r^b$, we have $\eta = \eta(r, x) \to 0$ as $r \to \infty$; hence, by Lemma 3.1(b),

$$P(|W| > r^b) \geq \frac{1}{2}(1 - \eta)^2 \exp\{r^{\rho - \epsilon} - (2r)^{\rho + \epsilon}\}$$

for $r > 0$ large enough. Therefore

$$\limsup_r \frac{\log\log P(|W| > r^b)^{-1}}{\log r} \leq \rho + \epsilon.$$

The result then follows by letting consecutively $b \to \rho - 1 - \epsilon$ and $\epsilon \to 0$.

Proposition 3.3. *(a) Let W be a random variable and Q be defined as in Definition 1.1. Then*

$$Q = \sup\{q \geq 0 : Ee^{|W|^q} < \infty\}.$$

(b) For all $q \in (0, \infty)$, defining R_q as in (1.4a), we have

$$R_q = \sup\{r \geq 0 : Ee^{r|W|^q} < \infty\}.$$

In particular, if $Q < \infty$, then

$$R \equiv R_Q = \sup\{r \geq 0 : Ee^{r|W|^Q} < \infty\} \ .$$

Proof. By considering $|W|$ instead of W, we can suppose $W \geq 0$ a.s. For part (a), let $a := \sup\{q \geq 0 : Ee^{W^q} < \infty\}$. An easy application of Markov's inequality gives $Q \geq a$. To prove that $a \geq Q$, let us notice that by the definition of Q, for all $0 < q < Q$, there is some A such that

$$P(W > x) \leq A \exp(-x^q)$$

for all $x > 0$, so by integration,

$$Ee^{W^{q'}} < \infty \text{ for all } 0 < q' < q.$$

This gives $a \geq q'$, and consequently $a \geq Q$.

For part (b), let $b := \sup\{r \geq 0 : Ee^{rW^q} < \infty\}$. An easy application of Markov's inequality for e^{rW^q} with $0 < r < b$ gives $R_q \geq b$ if $b > 0$. To prove that $b \geq R_q$, let us notice that by the definition (1.4a) of R_q, for all $0 < r < R_q$ there is some $B > 0$ such that for all $x > 0$,

$$P(W > x) \leq B \exp\{-rx^q\} \ ;$$

by integration, this gives the result desired.

Proposition 3.4. *For any random variable W, we have*

(a)
$$\mu = Q^{-1} \ \text{and} \ \underline{\mu} = \bar{Q}^{-1} \ ,$$

where for the last identity, we suppose additionally $Q > 0$ (or $\mu < \infty$).
(b) For all $q \in (0, \infty)$, let $R_q, \nu_q, \ldots$ be defined in (1.4a) and (1.4b). Then

$$(qR_q)(\nu_{1/q})^q e = 1 \ \text{and} \ (q\bar{R}_q)(\underline{\nu}_{1/q})^q e \geq 1.$$

In particular, if $0 < \mu < \infty$ (or $0 < Q < \infty$), then

$$e^\mu \nu = (QR)^{-1/Q} \ \text{and} \ e^\mu \underline{\nu} \geq (Q\bar{R})^{-1/Q} \ .$$

Proof. As in the preceding, by considering $|W|$ instead of W, we can suppose that $W \geq 0$ a.s.; we can also suppose that $\|W\|_\infty = \infty$ since otherwise $P(W > x) = 0$ for x large enough and then $\mu = \underline{\mu} = 0$ and $Q = \bar{Q} = \infty$.

(a1) We first prove that $\underline{\mu} \geq \bar{Q}^{-1}$ and $\mu \geq Q^{-1}$. Suppose that $\bar{Q} < \infty$. By the definition of $\bar{Q}$, for all $\epsilon < 0$ and all $x > 0$ large enough,

$$P(W > x) \geq \exp\{-x^{\bar{Q}+\epsilon}\} \ ; \tag{3.3a}$$

so by Markov's inequality,

$$EW^\lambda \geq x^\lambda P(W > x) \geq x^\lambda \exp\{-x^{\bar{Q}+\epsilon}\} \ .$$

Choosing $x = (\epsilon\lambda\log\lambda)^{1/(\bar{Q}+\epsilon)}$ and taking logarithm, we obtain

$$\underline{\mu} \geq (\bar{Q}+\epsilon)^{-1} - \epsilon.$$

This yields $\underline{\mu} \geq \bar{Q}^{-1}$. A very similar argument yields $\mu \geq Q^{-1}$.

(a2) We next prove that $Q \leq \mu^{-1}$ when $Q > 0$. For all $0 < q < Q$, there is some constant $C > 0$ such that

$$P(W > x) \leq C \exp\{-x^q\} \tag{3.4}$$

for all $x > 0$. By integration, this yields

$$EW^\lambda \leq C\frac{\lambda}{q}\Gamma\left(\frac{\lambda}{q}\right) \tag{3.5}$$

for all $\lambda \geq 1$. So by Stirling's formula, we get

$$\limsup_{\lambda\to\infty} \frac{\log EW^\lambda}{\lambda\log\lambda} \leq \frac{1}{q}. \tag{3.6}$$

Letting $q \to Q$ gives $\mu \leq 1/Q$.

(a3) We then prove that $\bar{\mu} \leq 1/\bar{Q}$ when $0 < \underline{\mu} \leq \bar{\mu} < \infty$. By (3.1), for all $x > 0$ and $\lambda > 0$, if $\eta = \eta(x,\lambda) := x^\lambda/EW^\lambda < 1$, then

$$P(W > x) = P(W^\lambda > \eta EW^\lambda) \geq (1 - \eta)^2 \left(EW^\lambda\right)^2/EW^{2\lambda}. \qquad (3.7)$$

By the definition of $\underline{\mu}$ and of $\bar{\mu}$, for all $0 < a < \underline{\mu}$ and all $\bar{\mu} < b < \infty$, we have

$$\exp\{a\lambda \log \lambda\} \leq EW^\lambda \leq \exp\{b\lambda \log \lambda\}, \qquad (3.8)$$

where $\lambda > 0$ is large enough. Let $x = \frac{1}{2}\lambda^a$. Then $\eta \to 0$ as $\lambda \to \infty$ and, by (3.7) and (3.8),

$$P(W > x) \geq 2 \exp\{3(a - b)\lambda \log \lambda\} \qquad (3.9)$$

for λ large enough. Letting $\lambda \to \infty$ yields $\bar{Q} \leq 1/a$, and letting $a \to \underline{\mu}$ yields $\bar{Q} \leq 1/\underline{\mu}$. This ends the proof of part (a).

(b1) From Proposition 3.3, R_q is the radius of convergence of the series

$$Ee^{rW^q} = \sum_{k=0}^\infty \frac{EW^{kq}}{k!} r^k;$$

so $R_q^{-1} = \limsup_k \left(\frac{EW^{kq}}{k!}\right)^{1/k}$. By the monotonicity of $\|W\|_p$ (in p) and Stirling's formula, we get

$$R_q^{-1/q} = \limsup_k \frac{\|W\|_{kq}}{(k/e)^{1/q}} = (eq)^{1/q}\nu_{1/q},$$

which establishes the first assertion of (b).

(b2) We now prove that $(q\bar{R}_q)^{-1/q} \leq e^{1/q}\underline{\nu}_{1/q}$ if $\bar{R}_q < \infty$ and if $EW^n < \infty$ for all $n > 0$. It is easy to see that for all $r > \bar{R}_q$ there is some constant $C > 0$ such that

$$P(W > x) \geq C \exp\{-rx^q\}$$

for all $x \geq 0$.

By integration, we have $EW^\lambda \geq C\frac{\lambda}{q}r^{-\lambda/q}\Gamma\left(\frac{\lambda}{q}\right)$ and then $\liminf_\lambda \frac{\|W\|_\lambda}{\lambda^{1/q}} \geq (eqr)^{-1/q}$. Letting $r \to \bar{R}_q$ yields $\underline{\nu}_{1/q} \geq (eq\bar{R}_q)^{-1/q}$.

For further studies, we need the following classical formulas about expressions of order and type of an entire function in terms of its Taylor coefficients.

Lemma 3.5. [Boas (1954), pp. 8–12] *Let $f(z) = \sum_{k=0}^\infty a_k z^k$ be an entire function and let $\rho, \underline{\rho}, \tau, \underline{\tau}$ be its order, lower order, type and lower type respectively. Then,*

(a)
$$\rho = \limsup_k \frac{k \log k}{-\log |a_k|} \quad \text{and} \quad \underline{\rho} \geq \liminf_k \frac{k \log k}{-\log |a_k|};$$

(b)
$$\rho\tau e = \limsup_k k \, |a_k|^{\rho/k} \quad \text{and} \quad \rho\underline{\tau}e \geq \liminf_k |a_k|^{\rho/k} \quad \text{if } 0 < \rho < \infty.$$

Proposition 3.6. (Calculations of ρ, τ, Q, and R in terms of moments) *Let W be a real r.v. with analytic or entire c.f., then*

(a) $$\rho = (1 - \mu)^{-1} \quad \text{and} \quad \underline{\rho} \geq (1 - \underline{\mu})^{-1};$$

(b) $$(\rho\tau)^{1/\rho} = e^\mu \nu \quad \text{and} \quad (\underline{\rho}\underline{\tau})^{1/\underline{\rho}} \geq e^{\underline{\mu}} \underline{\nu} \qquad \text{if } 1 < \rho < \infty.$$

The proof is straightforward using Lemma 3.5 and Stirling's formula.

Lemma 3.7. *If $1 < \rho < \infty$ and $0 \leq \underline{\tau} < \tau \leq \infty$, then*

$$\bar{R} \leq (2\rho\tau - 2\underline{\tau})\underline{\tau}^{-Q}.$$

The proof uses Lemma 3.1 for a convenient r and estimates of $M(r)$ given by the definitions of τ and $\underline{\tau}$.

Corollary. *If $1 < \rho < \infty$ and $0 < \underline{\tau} \leq \tau < \infty$, then $0 < R \leq \bar{R} < \infty$.*

The fact that $0 < \underline{\tau} = \tau < \infty$ is a little more difficult to prove directly, and is contained in the following lemma.

Lemma 3.8. *If $1 < \rho < \infty$ and $0 < \underline{\tau} \leq \tau < \infty$, then*

$$\bar{R}/R \leq (\bar{\lambda}/\lambda)^Q,$$

where $0 < \lambda \leq \rho\underline{\tau} \leq \rho\tau \leq \bar{\lambda} < \infty$ are the two positives solutions of the equation

$$x - Rx^Q = \underline{\tau} ,$$

and $\lambda = \bar{\lambda}$ if and only if $\underline{\tau} = \tau$.

Proof. By Lemma 1.2, we can suppose that $W \geq 0$ a.s. So by Davies' (1976) theorem,
$$\bar{R}/R \leq (\bar{r}/r)^{\rho/(\rho-1)},$$
$0 < r \leq \bar{r} < \infty$ being the two positive solutions of $f(y) = 0$, where

$$f(y) = (\rho - 1)(\tau/\bar{\tau})^{1/(\rho-1)} - \rho y + 1;$$

the equation $f(y) = 0$ does have exactly two positive solutions $0 < r \leq 1 \leq \bar{\tau}/\tau \leq \bar{r} < \infty$ and $r = \bar{r}$ if and only if $\tau = \bar{\tau}$ because $f(0) = 1 > 0$, $f(1) \leq 0$, $f'(\bar{\tau}/\tau) = 0$, $f(+\infty) = +\infty$ *and* $f''(y) > 0$ *if* $0 < y < \infty$. The conclusion then follows since by (1.6d),
$$\bar{\tau}f(y) = R(\rho\bar{\tau}y)^Q - (\rho\bar{\tau}y) + \bar{\tau}.$$

We are now in the position to end the proof of Theorems 1.1 and 1.2. Theorem 1.1 comes directly from Proposition 3.3. For Theorem 1.2, we notice that part (a1) is Proposition 3.4(a); part (a2) comes from Proposition 3.4(b). Part (b0) is a consequence of (b1), and (b1) is a combination of (a1), Propositions 3.6 and 3.2, and the Corollary of Lemma 3.7. Part (b2) comes from Propositions 3.4(b) and 3.6(b). Part (b3) is a consequence of (b2) and Lemma 3.8.

4   Proof of Theorem 2.1

In this section, we assume always the conditions (2.2), (2.3), (2.5) and (2.6), and let W and $\beta \in [0,1]$ be defined in (2.4) and (2.7). For $0 \le q < \infty$, R_q, $\bar{R}_q$, ν_q and $\underline{\nu}_q$ are defined in (1.4).

The following result generalizes a result of Graf et al.(1988, Th. 2.5, p.14); we recall that $R_{1/q} > 0$ if and only if $\nu_q < \infty$.

Theorem 4.1. *(a) If $\|N\|_\infty < \infty$, then $\nu_1 \equiv \limsup_k \|W\|_k/k < \infty$ and, for all $b \in (0,1)$,*

$$\nu_b \equiv \limsup_k \|W\|_k/k^b < \infty \tag{4.1}$$

if and only if

$$S\big(1/(1-b)\big) \le 1 \ a.s. \tag{4.2}$$

(b) If $\|N\|_\infty = \infty$, then for all $b \in (0,1)$, (4.1) implies (4.2).

Proof. (i) We first prove that if $\|N\|_\infty < \infty$, then $R_1 > 0$ and, for all $b \in (0,1)$, (4.2) implies (4.1). Fix $b \in (0,1]$ and, if $b \ne 1$ we suppose that (4.2) holds. Since W is solution of (2.1), we have for $n \ge 1$,

$$E[W^k \mid N = n] = E\Big(\sum_{i=1}^n A_i^k \mid N = n\Big)EW^k$$

$$+ \sum_n{}' \frac{k!}{k_1!\ldots k_n!} E\Big(\prod_{i=1}^n A_i^{k_i} \mid N = n\Big) \prod_{i=1}^n E[W^{k_i}] \tag{4.3}$$

where $\sum_n{}'$ means a sum extended over all $(k_1,\ldots,k_n)$ such that $0 \le k_i \le k-1$ for all $i = 1,\ldots,n$ and $k_1+\ldots+k_n = k$. By considering the maximum of $\prod_{i=1}^n x_i^{(1-b)k_i}$ on $\Delta = \{(x_1,\ldots,x_n) \in [0,1]^n : \sum_{i=1}^n x_i \le 1\}$, it is easy to see that if $b \in (0,1)$ and if $\sum_{i=1}^n A_i^{1/(1-b)} \le 1$, then

$$\prod_{i=1}^n A_i^{k_i} \le \Big(\frac{k_i}{k}\Big)^{(1-b)k_i} ; \tag{4.4}$$

if $b = 1$ it is obvious. Thus if $b \in (0,1]$, by(4.2)-(4.4), we have

$$E[W^k \mid N = n] = E\Big(\sum_{i=1}^n A_i^k \mid N = n\Big)EW^k + \sum_n{}' \frac{k!}{k_1!\ldots k_n!} \prod_{i=1}^n \Big(\frac{k_i}{k}\Big)^{(1-b)k_i} E[W^{k_i}]$$

for $n \ge 1$ and $k \ge 2$.

Therefore, letting $t_p = \dfrac{EW^p}{p!} p^{(1-b)p}$ for $p \ge 1$ and $t_0 = 1$, we get

$$t_k \le \frac{1}{1-\rho(k)} E \sum_N{}' \prod_{i=1}^N t_{k_i} . \tag{4.5}$$

Since $\sum_N' \prod_{i=1}^N$ is an increasing function of N, we have, for $n := \|N\|_\infty$,

$$t_k \leq c \sum_n' \prod_{i=1}^n t_{k_i} \;, \qquad k \geq 2, \tag{4.6}$$

where $c := \sup_{k \geq 2} 1/[1 - \rho(k)] = 1/[1 - \rho(2)] > 0$. By (4.6) and Lemma 2.6 of Graf et al. (1988) we obtain $\limsup_k t_k^{1/k} < \infty$. This implies $\limsup_k e\|W\|_k/k^b < \infty$.

(ii) We now prove that if

$$P(S(1/(1-b)) > 1) > 0 \;, \tag{4.7}$$

then $\nu_b = \infty$. We remark that the latter holds if

$$\liminf_k s_k^{1/k} = \infty \;, \tag{4.8}$$

$$\text{where} \qquad s_k = \frac{EW^k}{(k!)^b} \;, \quad k \geq 0. \tag{4.9}$$

We shall prove that (4.7) implies (4.8) without assuming $\|N\|_\infty < \infty$. From (4.3) we have

$$E[W^k][1 - E(\sum_{i=1}^N A_i^k)] = E\{\sum_N' \frac{k!}{k_1! \ldots k_N!} \prod_{i=1}^N A_i^{k_i} \prod_{i=1}^N EW^{k_i}\}. \tag{4.10}$$

Suppose that $s_j \geq r^j$ for some $r > 0$ and all $j < k$, we shall see that $s_k \geq r^k$ if k is large enough. In fact, from (4.10) and the inequality $\sum x_i^{1-b} \geq (\sum x_i)^{1-b}$, we get

$$EW^k(1 - \rho(k)) \geq E\{\sum_N' \frac{k!}{k_1! \ldots k_N!} \prod_{i=1}^N A_i^{k_i} r^k (k_1! \ldots k_N!)^b\}$$

$$\geq E\{(\sum_N' \frac{k!}{k_1! \ldots k_N!} \prod_{i=1}^N A_i^{k_i/(1-b)})^{1-b}\} r^k (k!)^b \;.$$

Therefore, remarking that

$$\sum_N' \frac{k!}{k_1! \ldots k_N!} \prod_{i=1}^N A_i^{k_i/(1-b)} = (\sum_{i=1}^n A_i^{1/(1-b)})^k - \sum_{i=1}^n A_i^{k/(1-b)}$$

and that $(x-y)^{1-b} \geq x^{1-b} - y^{1-b}$ if $x \geq y \geq 0$, we obtain

$$EW^k(1 - \rho(k)) \geq E\left[(\sum_{i=1}^N A_i^{1/(1-b)})^{k(1-b)} - (\sum_{i=1}^N A_i^{k/(1-b)})^{1-b}\right] r^k (k!)^b \;.$$

Again by the inequality $\sum x_i^{1-b} \geq (\sum x_i)^{1-b}$, this yields

$$\frac{EW^k}{k!^b} \geq \frac{E\left([S(\frac{1}{1-b})]^{k(1-b)}\right) - \rho(k)}{1 - \rho(k)} r^k \;. \tag{4.11}$$

From (4.7), we can choose $a > 1$ such that $c := P\big[S(1/(1 - b)) > a\big] > 0$, so $E\big(\big[S(1/(1 - b))\big]^{k(1-b)}\big) \geq ca^{k(1-b)}$. Choose $k_0 > 0$ such that $ca^{k_0(1-b)} > 1$. Put $r_0 = \min_{0 \leq i \leq k_0} s_i^{1/i} > 0$, then (4.11) implies $s_k^{1/k} \geq r_0$ for all $k \geq k_0$. Thus $\liminf_k s_k^{1/k} = \liminf\big(\frac{EW^k}{k!^b}\big)^{1/k} > 0$. Since we can choose $b' > b$ such that (4.7) holds for b', we have actually that $\liminf\big(\frac{EW^k}{k!^b}\big)^{1/k} = \infty$. This ends the proof of both (a) and (b).

Corollary 4.1. *Let $b \in (0, 1)$. If $P(S(1/(1 - b)) > 1) > 0$, then*

$$\nu_b = \underline{\nu}_b = \lim_k \|W\|_k/k^b = \infty. \tag{4.12}$$

Proof. It comes from (4.8) and Stirling's formula.

Theorem 4.2. *(a) If $\|N\|_\infty < \infty$, then $\mu = \underline{\mu} = \beta$ and $\nu = \nu_\beta < \infty$; (b) if $\|N\|_\infty = \infty$, then $\mu \geq \underline{\mu} \geq \beta$.*

Proof. We first prove that in both cases $\|N\|_\infty < \infty$ and $\|N\|_\infty = \infty$, $\underline{\mu} \geq \beta$. If $\beta = 0$, then there is nothing to prove; if $\beta > 0$ and if $0 < b < \beta$, then $P(S((1 - b)^{-1}) > 1) > 0$. So, by Corollary 4.1, $\|W\|_k \geq k^b$ for all k large enough. This yields $\underline{\mu} = \liminf_k \log \|W\|_k/ \log k \geq b$. Letting $b \to \beta$ gives $\underline{\mu} \geq \beta$.

We now assume $\|N\|_\infty < \infty$. If $\beta = 1$, then $\nu_\beta = \nu_1 < \infty$ by Theorem 4.2. If $\beta < 1$, then $S(1/(1 - b)) \leq 1$ a.s. for all $b \in (\beta, 1)$; letting $b \to \beta$ gives $S(1/(1 - \beta)) \leq 1$ a.s. So Theorem 4.1 applies again, yielding that $\nu_\beta < \infty$. Since $\mu = \inf\{b > 0 : \nu_b < \infty\}$, it follows that $\mu \leq \beta$; combining this with $\underline{\mu} \geq \beta$ gives $\mu = \underline{\mu} = \beta$.

To ensure $R_{1/\beta} < \infty$ if N is constant, Graf et al. (1988) have given a "corner" condition (Th. 2.11 and Cor. 2.12-14, pp. 30-37), which seems rather complicated. Here we give a simple result covering all their examples and all those of Falconer (1986). [See Liu (1993).]

Theorem 4.3. *Let $b \in [0, \infty)$. If there exists $n > 1$ such that*

$$\prod_{i=1}^{\infty} \frac{\|S(1)1_{\{N=n\}}\|_{n^i}}{n^b} > 0, \tag{4.13a}$$

then $\underline{\nu}_b \equiv \liminf_k \frac{\|W\|_k}{k^b} > 0\big($ and so $R_{1/b} < \infty$ if $b > 0\big)$.

Proof. Let $n > 1$ be such that (4.13a) holds. From (4.3), we have, for all $k \geq 2$,

$$E[W^k \mid N = n] \geq E\Big(\sum_{i=1}^{n} A_i^k \mid N = n\Big)EW^k$$

$$+ \big(\inf \prod_{i=1}^{\infty} E[W^{k_i}]\big) \sum_n{}' \frac{k!}{k_1! \dots k_n!} E\big[\prod_{i=1}^{n} A_i^{k_i} \mid N = n\big],$$

where the infimum is taken over all the $(k_1, \ldots, k_n)$ of the sum. If $k = n\tilde{k}$, it is $(EW^{\tilde{k}})^n$. Therefore, we get consecutively

$$E[W^k \mid N = n] \geq (EW^{\tilde{k}})^n E\Big[\Big(\sum_{i=1}^{n} A_i\Big)^{n\tilde{k}} \mid N = n\Big].$$

$$E[W^{n\tilde{k}}] \geq \big(E[W^{\tilde{k}}]\big)^n E\big[S(1)^{n\tilde{k}} 1_{\{N=n\}}\big], \quad \text{and} \quad \|W\|_{n\tilde{k}} \geq \|W\|_{\tilde{k}} \|S(1) 1_{\{N=n\}}\|_{n\tilde{k}}.$$

Choosing $\tilde{k} = n^{r-1}$ $(r > 0)$ and iterating, we see that

$$\frac{\|W\|_{n^r}}{n^{rb}} \geq \prod_{j=1}^{r} \frac{\|S(1) 1_{\{N=n\}}\|_{n^j}}{n^b}.$$

Letting $r \to \infty$ gives $\liminf_r \frac{\|W\|_{n^r}}{n^{rb}} > 0$. Since $\|W\|_k$ is increasing, it follows that $\nu_b \geq \underline{\nu}_b \equiv \liminf_k \frac{\|W\|_k}{k^b} > 0$.

Corollary 4.3. *If for some constant* $n > 1$,

$$\beta = log\|S(1) 1_{\{N=n\}}\|_\infty / \log n \tag{4.13b}$$

and

$$\prod_{i=1}^{\infty} \frac{\|S(1) 1_{\{N=n\}}\|_{n^i}}{\|S(1) 1_{\{N=n\}}\|_\infty} > 0 \tag{4.13c}$$

then $\underline{\nu}_\beta \equiv \liminf_k \frac{\|W\|_k}{k^\beta} > 0$.

Proof. It suffices to remark that (4.13a) holds for $b = \beta$ if and only if (4.13b) and (4.13c) hold.

Lemma 4.4. *Let* $n \geq 1$ *be a constant integer and let* $(X_1, \ldots, X_n)$ *be a sequence of i.i.d.r.v.'s with* $P(X_1 \geq 0) = 1$ *and* $\|X_1\|_\infty < \infty$. *Put* $S_n = X_1 + \ldots + X_n$. *Then for all* $p \in \{2, 3, \ldots\}$,

$$\prod_{i=1}^{\infty} \frac{\|S_n\|_{p^i}}{\|S_n\|_\infty} > 0 \iff \prod_{i=1}^{\infty} \frac{\|X_1\|_{p^i}}{\|X_1\|_\infty} > 0.$$

Proof. The implication "$\Rightarrow$" is easy by the triangular inequality for $\|\cdot\|_{p^i}$ and the fact that $\|S_n\|_\infty = n\|X_1\|_\infty$. For the converse, we remark that for all $k=1,2,\ldots$

$$E[S_n^k] = \sum \frac{k!}{k_1! \ldots k_n!} E\Big(\prod_{i=1}^{n} X_i^{k_i}\Big) \geq n^k \inf E\Big(\prod_{i=1}^{n} X_i^{k_i}\Big),$$

where the sum and the infimum are taken over all $(k_1, \ldots, k_n) \in \mathbf{N}^n$ such that $k_1 + \ldots + k_n = k$; the infimum is equal to $[E(X_1^{\tilde{k}})]^p$ if $k = \tilde{k}p$. Therefore

$$\|S_n\|_{\tilde{k}p} \geq n\|X_1\|_{\tilde{k}}, \qquad \tilde{k} = 1, 2, \ldots$$

So $\|S_n\|_{p^i} \geq n\|X_1\|_{p^{i-1}}$ and the conclusion follows.

Lemma 4.5. *Assume $\|N\|_\infty < \infty$ and $\beta < 1$. If for some $0 < d < 1$ and $1 < p < d^{-1/(1-\beta)}$,*

$$\prod_{k=1}^{\infty} \| \sum_{i=1}^{N} A_i^{1/(1-\beta)} 1_{\{\sup_{1 \le i \le N} \le d\}} \|_{p^k} > 0, \tag{4.14}$$

then $\underline{\nu}_\beta \equiv \liminf_k \frac{\|W\|_k}{k^\beta} > 0$ (and so $R_{1/\beta} < \infty$).

Proof. We can suppose that $p \notin \mathbf{N}$, since otherwise we can choose $\tilde{p} \notin \mathbf{N}$ with $p < \tilde{p} < d^{-1/(1-\beta)}$ and consider $\tilde{p}$ instead of p. Moreover, by completing the A_i's with 0 if necessary, we can also suppose that $N = n$ is a.s. a constant large enouph such that $n > p$. Put $\delta = 1 - d^{-1/(1-\beta)}$ and $a = 1/p$. Then $a > \max(1/n, 1-\delta)$, $a \notin \{1/\nu : \nu = 1, \dots, n-1\}$ and, by (4.14), the condition (2.92) of Graf et al.(1988) holds. Therefore, by the proofs of their Theorem 2.11 and Corollary 2.12, we get

$$\liminf_{k \to \infty} \sqrt[k]{s_k(1/\beta)} > 0, \quad \text{where} \quad s_k(1/\beta) = EW^k/\Gamma(\beta k + 1). \tag{4.15}$$

In fact, in their proofs, (2.92) implies (2.100) which is nothing but (2.77); and (2.77) implies (2.82), (2.88) and (2.89); combining (2.82), (2.88) and (2.89) gives our (4.15). By Stirling's formula, (4.15) implies $\underline{\nu}_\beta > 0$. (So $\nu_\beta > 0$ and $R_{1\beta} < \infty$ by (1.4c).)

Lemma 4.6. *Let X be a r.v. with values in $[0, 1]$. If for some constants $c > 0$, $a \ge 0$ and all $x > 0$ small enough,*

$$P(X > 1 - x) \ge cx^a, \tag{4.16}$$

then for some constant $\tilde{c} > 0$ and all sufficiently large $k > 0$,

$$EX^k \ge \tilde{c}/k^{a+1} . \tag{4.17}$$

Consequently, for all $p > 1$,

$$\prod_{i=1}^{\infty} \|X\|_{p^i} > 0. \tag{4.18}$$

Proof. Let $0 < \delta < 1$ be such that $P(X > y) \ge c(1 - y)^a$ for all $0 < y < \delta$. Therefore for all $k > 1$,

$$EX^k = \int_0^1 ky^{k-1} P(X > y)\, dy \ge c \int_\delta^1 ky^{k-1}(1 - y)^a\, dy \ge cB(k, a+1) - c\delta^k,$$

where $B(k, a+1) = \int_0^1 y^{k-1}(1 - y)^a\, dy = \Gamma(k)\Gamma(a+1)/\Gamma(k+a+1)$. So (4.17) holds by Stirling's formula. (4.18) follows from (4.17).

Lemma 4.7. *(a) Let X be a r.v. with values in $[0, 1]$. If X has a density function $d(x)$ with respect to the Lebesgue measure on $[0, 1]$ such that, for some constants $c > 0$, $a \ge 0$, $\delta \in (0, 1)$ and all $0 < x < \delta$,*

$$d(1 - x) \ge cx^a,$$

then for all $0 < x < \delta$,

$$P(X > 1 - x) \geq \frac{cx^{a+1}}{a+1}. \qquad (4.19a)$$

(b) Let $n \geq 1$ be a constant integer and let $(X_1, \ldots, X_n)$ be a random variable with values in $\Delta := \{(x_1, \ldots, x_n) \in [0,1]^n : x_1 + \ldots + x_n \leq 1\}$. Suppose that $(X_1, \ldots, X_n)$ has a density $d(x_1, \ldots, x_n)$ (with respect to the Lebesgue measure) on Δ which satisfies the following property: there exist a point $(x_1^0, \ldots, x_n^0) \in \Delta$ with $x_i^0 > 0$ and $x_1^0 + \ldots + x_n^0 = 1$ such that, for some constants $c > 0$, $a \geq 0$ and for all $0 \leq x_i < x_i^0 (1 \leq i \leq n)$ sufficiently near to x_i^o,

$$d(x_1, \ldots, x_n) \geq c[1 - (x_1 + \ldots + x_n)]^a.$$

Then for some constant $0 < \delta < 1$ and for all $x > 0$ sufficiently small,

$$P\{\sum_{i=1}^{n} X_i > 1 - x \text{ and } A_i \leq \delta \text{ for all } 1 \leq i \leq N\} \geq \frac{cx^{a+n}}{(a+1)\cdots(a+n)}. \qquad (4.19b)$$

Proof. Part (a) is easy, and can be regarded as a special case of part (b) with $n = 1$. Suppose that $n \geq 2$. Let $\epsilon > 0$ be sufficiently small such that for all $x_j \in (x_j^0 - \epsilon, x_j^0), 1 \leq j \leq n$, $d(x_1, \ldots, x_n) \geq [1 - (x_1 + \ldots + x_n)]^a$. Therefore for $\delta := \max_{1 \leq i \leq n} x_i^0 \in (0,1)$ and all $0 < x < \epsilon$, the left member of (4.19b) is not less than

$$I := c \int_{x_1^0 - \epsilon}^{x_1^0} dx_1 \cdots \int_{x_n^0 - \epsilon}^{x_n^0} [1 - (x_1 + \ldots + x_n)]^a 1_{\{x_1 + \ldots + x_n > 1 - x\}} \, dx_n.$$

Remarking that $x_1 + \ldots + x_{n-1} + x_n^0 - \epsilon \leq 1 - \epsilon < 1 - x$, we have

$$\int_{x_n^0 - \epsilon}^{x_n^0} 1_{\{x_1 + \ldots + x_n > 1 - x\}} \, dx_n \geq [t_{n-1} - (1-x)] 1_{\{t_{n-1} > 1 - x\}}.$$

where $t_{n-1} = x_1 + \ldots + x_{n-1} + x_n^0$; so

$$I \geq c \int_{x_1^0 - \epsilon}^{x_1^0} dx_1 \cdots \int_{x_{n-1}^0 - \epsilon}^{x_{n-1}^0} [1 - t_{n-1}]^a [t_{n-1} - (1-x)] 1_{\{t_{n-1} > 1 - x\}} \, dx_{n-1}.$$

Proceeding in a similar way, we get

$$I \geq \frac{c}{(n-1)!} \int_{x_1^0 - \epsilon}^{x_1^0} [1 - t_1]^a [t_1 - (1-x)]^{n-1} 1_{\{t_1 > 1 - x\}} \, dx_1,$$

where $t_1 = x_1 + x_2^0 + \ldots + x_n^0$. By changing variables $t = 1 - t_1$ and $u = t/x$, we obtain

$$I \geq \frac{c}{(n-1)!} \int_0^x t^a (x - t)^{n-1} \, dt = \frac{c}{(n-1)!} x^{a+n} B(a+1, n),$$

where $B(a+1, n) = \int_0^1 u^a (1-u)^{n-1} \, du = \Gamma(a+1)\Gamma(n)/\Gamma(a+1+n)$. So (4.19b) holds.

Lemma 4.8. *If* $0 < \beta < 1$ *and if for all* $1 \le i \le N$, T_i *takes only finitely many values a.s., then*

$$P[S(1/(1-\beta)) = 1 \text{ and } \tilde{N} \ge 2] > 0.$$

Proof. Suppose that $P[S(1/(1-\beta)) = 1$ and $\tilde{N} \ge 2] = 0$. Then $S(1/(1-\beta)) < 1$ a.s. on $\{\tilde{N} \ge 2\}$, and so for some $0 < \tilde{\beta} < \beta$, $S(1/(1-\tilde{\beta})) < 1$ a.s. on $\{\tilde{N} \ge 2\}$. As on $\{\tilde{N} \le 1\}$, $S(1/(1-\tilde{\beta})) = \sup A_i^{1/(1-\beta)} \le 1$ a.s., it follows that $S(1/(1-\tilde{\beta})) < 1$ a.s. This is a contradiction with the definition of β; so the proof is completed.

The proof of Theorem 2.1 is then a simple combination of the results of this section and Theorems 1.1 and 1.2, as is shown in the following. (2.8b) comes from Theorem 4.2; (2.8a) and (2.8c) follow from (2.8b) andTheorem 1.2. If (2.9b) holds, then (2.9b') follows by Lemma 4.6; so (2.9a) and (2.9b) imply $\underline{\nu}_\beta > 0$ by Corollary 4.3. If (2.9c) holds, then again by Lemma 4.6, (2.9c') holds since the left termin (2.9c) can be rewritten as $P[S(1/(1-\beta))1_{\{\sup_{1\le i \le N} A_i \le \delta\}} > 1 - x]$ for all $0 < x < 1$. So by Lemma 4.5, we have again $\underline{\nu}_\beta > 0$. Combining this with (2.8b), we obtain (2.10b). Since (2.10b) is equivalent to both (2.10a) and (2.10c) by Theorem 1.3(b1), the proof of Theorem 2.1 is completed.

In Remark 2.1, part (1) is seen by the proof above. For part (2), we note that (a) implies (b) by Lemma 4.8; (b) and (c) are equivalent each other; if (c) holds, then for some $0 < \delta < 1$, $P[S(1/(1-\beta)) = 1$ and $\sup_{1\le i \le N} A_i \le \delta\}] > 0$, so (2.9c) holds with $a = 0$. If (d) holds, then (2.9c) follows from Lemma 4.7. Corollary 2.1 comes from Remark 2.1(1), Lemma 4.4 and Lemma 4.6.

Acknowledgement. The author is very grateful to Professors Yves Guivarc'h and Alain Rouault for many discussions, remarks and help.

References

Athreya, K.B. (1971): Galton-Watson branching processes. *J. Appl. Prob.* **8**, 589–598.

Athreya, K.B. and Ney, P.E. (1972): *Branching processes*. Berlin: Springer.

Ben Nasr, F. (1987): Mesures aléatoires de Mandelbrot associées à dessubstitutions. *CRAS, Sér. I* **304**, 255–258.

Biggins, J.D. (1977): Martingale convergence in the branching random walk. *J. Appl. Prob.* **14**, 25–37.

Biggins, J.D. and Bingham, N.H. (1993): Large deviations in the supercritical branching process. *Adv. Appl. Prob.* **25**, 757–772.

Bingham, N.H. and Doney, R.A. (1974): Asymptotic properties of supercritical branching processes I: The Galton-Watson process. *Adv. Appl. Prob.* **6**,711–731.

Bingham, N.H. and Doney, R.A. (1975): Asymptotic properties of supercritical branching processes II: Crump-Mode and Jirina processes. *Adv. Appl. Prob.* **7**, 66–82.

Bingham, N.H., Goldie, C.M. and Teugels, J.L. (1987): *Regular variation*. Cambridge: Cambridge University Press.

Boas, R.P. (1954): *Entire functions*. New York: Academic Press.

Chauvin, B. and Rouault, A. (1996): Boltzmann-Gibbs weigths in the branching random walk. In *Classical and Modern Branching processes*, eds K.B. Athreya, P. Jagers, IMA Proceedings **84**, 41–50. Springer-Verlag, New York.

Collet, P. and Koukiou, F. (1992): Large deviations for multiplicative chaos. *Comm. Math. Phys.* **147**, 329–342.

Crump, K. and Mode, C.J. (1968–1969): A general age-dependent branching process. *J. Math.Anal. Appl.* **24**, 497–508 and 25, 8–17.

Davies, L. (1976): Tail probabilities for positive random variables with entire characteristic functions of very regular growth. *Z. Angew. Math.* **56**, 334–336.

Dembo and Zeitouni (1993): *Large Deviations Techniques and applications*. Ed. Jones and Bartlett.

Doney, RÅ. (1972): A limit theorem for a class of supercritical branching processes. *J. Appl. Prob.* **9**, 707–724.

Doney, R.A. (1973): On a functional equation for general branching processes. *J. Appl. Prob.* **10**, 497–508.

Dugué, D. (1939): Sur quelques propriétés analytiques des fonctions caractéristiques. *CRAS* **208**, 1778–1780.

Durrett, R. and Liggett, T. (1983): Fixed points of the smoothing transformation. *Z. Wahrsch. verw. Gebiete* **64**, 275–301.

Falconer, K.J. (1986): Random fractals. *Math. Proc. Camb. Phil. Soc.* **100**, 559–582.

Falconer, K.J. (1987): Cut set sums and tree processes. *Proc. Amer. Math. Soc. (2)* **101**, 337–346.

Franchi, J. (1993): Chaos multiplicatif : un traitement simple et complet de la fonction de partition. Séminaire de Probabilités XXIX, 194–201, Lecture Notes in Mathematics, Berlin: Springer.

Graf, S., Mauldin, R.D. and Williams, S.C. (1988): The exact Hausdorff dimensionin random recursive constructions. *Mem. Amer. Math. Soc.* **71**, 381.

Guivarc'h, Y. (1990): Sur une extension de la notion de loi semi-stable. *Ann. IHP* **26**, 261–285.

Harris, T.E. (1948): Branching processes, *Ann. Math. Stat.* **19**, 474–494.

Holley, R. and Liggett, T. (1981): Generalized potlatch and smoothing processes. *Z. Wahrsch. verw. Gebiete* **55**, 165–195.

Holley, R. and Waymire, E.C. (1992): Multifractal dimensions and scaling exponents for strongly bounded cascades. *Ann. Appl. Prob.* **2**, 819–845.

Kahane, J.P. (1985): *Some Random Series of Functions*. 2nd. ed. Cambridge: Cambridge university press.

Kahane, J.P. (1987): Multiplications aléatoires et dimension de Hausdorff. *Ann. IHP, Sup. au no. 2*, vol. 23, 289–296.

Kahane, J.P. and Peyrière (1976): Sur certaines martingales de Benoit Mandelbrot. *Adv. Math.* **22**, 131–145.

Kasahara,Yuji (1977): Tauberian theorems of exponential type. *J. Math.Kyoto Univ.* **18**, 209–219.

Kesten, H. and Stigum, B.P. (1966): A limit theorem for multidimensional Galton-Watson processes. *Ann.Math.Statist.* **37**, 1211–1223.

Liu, Q. (1993): *Sur quelques problèmes à propos des processus de branchement, des flots dans les réseaux et des mesures de Hausdorff associées.* Thèse, Université Paris 6.

Liu, Q. (1994): *Sur une équation fonctionnelle et ses applications : une extension du théorème de Kesten-Stigum concernant des processus de branchement.* Prépublication Rennes 1.

Liu, Q. (1995): *Fixed points of a generalized smoothing transformation andapplications to branching processes.* Prépublication, Rennes 1.

Liu, Q. (1996): The exact Hausdorff dimension of a branching set. To appear in *Prob. Th. and Rel. Fields.*

Lukacs, E. (1964): *Fonctions caractéristiques.* Dunod, Paris.

Mandelbrot, B. (1974): Multiplications aléatoires et distributions invariantes par moyenne pondérée aléatoire. *CRAS Paris* **278**, 325–346 et 355–358.

Mauldin, R.A. and Williams, S.C. (1986): Random constructions, asymptotic geometric and topological properties. *Trans. Amer. Math. Soc.* **295**, 325–346.

Ramachandran, B. (1962): On the order and the type of entire characteristic functions.*Ann. Math. Stat.* **33**, 1238–1255.

Seneta, E. (1968): On recent theorems concerning the supercritical Galton-Watson process. *Ann. Math. Stat.* **39**, 2098–2102.

Seneta, E. (1969): Functional equations and the Galton-Watson process. *Adv. Appl. Prob.* **1**, 1–42.

Waymire, E.C. and Williams, S.C. (1995): Multiplicative cascades: dimension spectra and dependence. *J. of Fourier Analysis and Appl.*, Kahane Special Issue.

Quansheng Liu
Institut Mathématique de Rennes
Université de Rennes 1
Campus de Beaulieu
35042 Rennes, France

liu@levy.univ-rennes1.fr

Progress in Probability, Vol. 40
© 1996 Birkhäuser Verlag Basel/Switzerland

Probabilistic Aspects of Infinite Trees and Some Applications

RUSSELL LYONS

1991 Mathematics Subject Classification: Primary 60J15, 60J80, 60K35, 05C05. Secondary 60D05, 82A43, 05C80, 60B15

Key words and phrases: Galton-Watson, branching processes, Hausdorff dimension, capacity, Cayley graph, finitely generated group, electrical networks, random walks, percolation, fractals

Abstract. We begin by using flows to assign a positive real number, called the branching number, to an arbitrary (irregular) infinite locally finite tree. The branching number represents the "average" number of branches per vertex and is the exponential of the dimension of (the boundary of) the tree, as introduced by Furstenberg. There are many senses in which this branching number is an average. We discuss some based variously on electrical networks, random walks, percolation, or tree-indexed (branching) random walks. A refinement of the notion of branching number uses ideas of potential theory. This creates quite precise connections among probabilistic processes on trees.

For applications, we consider the structure of the family trees of branching processes, the Hausdorff dimension and capacities of possibly random fractals, and random walks on Cayley graphs of infinite but finitely generated groups.

Résumé. Nous utilisons d'abord les flots pour associer un réel positif appelé le nombre de branchement à un arbre arbitraire (non régulier) infini, localement fini. Le nombre de branchement représente le nombre "moyen" de branches par sommet et c'est l'exponentielle de la dimension de (la frontière de) l'arbre, déjà introduit par Furstenberg. Ce nombre de branchement est une moyenne en de nombreux sens. Nous en discutons certains, issus des réseaux électriques, des marches aléatoires, de la percolation ou des marches aléatoires (avec branchement) indexées par un arbre. Un raffinement de la notion de nombre de branchement utilise les idées de la théorie du potentiel. Ceci crée des connections tout à fait précises entre les processus probabilistes sur les arbres.

Comme applications, nous considérons la structure des arbres généalogiques des processus de branchement, la dimension de Hausdorff et les capacités de fractals éventuellement aléatoires, et les marches aléatoires sur des graphes de Cayley de groupes infinis finiment engendrés.

1 Branching Number

How do we assign an average branching number to an arbitrary infinite locally finite tree? If the tree is a binary tree, as in Figure 1, then clearly the answer will be "2". But in the general case, since the tree is infinite, no straight average is available. We must take some kind of limit or use some other procedure.

One simple idea is as follows. Let S_n be the set of vertices at distance n from the root, 0. Define the **growth rate** of the tree to be

$$\operatorname{gr} T := \liminf_{n\to\infty} |S_n|^{1/n} .$$

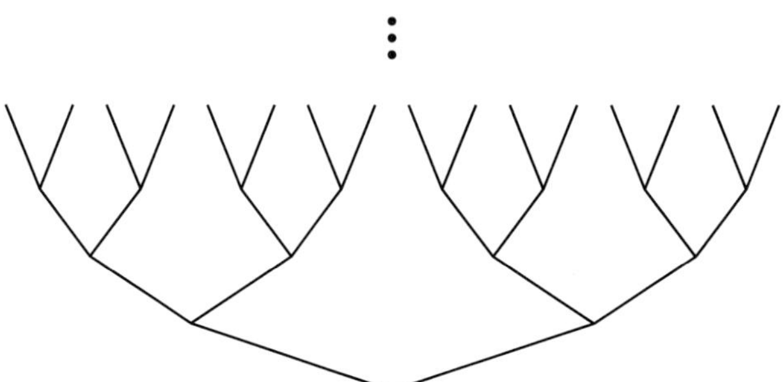

Figure 1.   The binary tree.

This certainly will give the number "2" to the binary tree. However, notice that it barely accounts for the structure of the tree: only $|S_n|$ matters, not how the vertices at different levels are connected to each other.

Thus, we shall use a different approach. Consider the tree as a network of pipes and imagine water entering the network at the root. However much water enters a pipe leaves at the other end and splits up among the outgoing pipes (edges). Consider the following sort of restriction: Given $\lambda \geq 1$, suppose that the amount of water that can flow through an edge at distance n from 0 is only λ^{-n}. If λ is too big, then perhaps no water can flow. In fact, consider the binary tree. A moment's thought shows that water can still flow throughout the tree provided that $\lambda \leq 2$, but that as soon as $\lambda > 2$, then no water at all can flow. Obviously, this critical value of 2 for λ is the same as the branching number of the binary tree. So let us make a general definition: the **branching number** of a tree T is the supremum of those λ that admit a positive amount of water to flow through T; denote this critical value of λ by $\operatorname{br} T$. As we shall see, this definition is the exponential of what Furstenberg (1970) called the "dimension" of a tree, which is the Hausdorff dimension of its boundary.

It is not hard to check that $\operatorname{br} T$ is related to $\operatorname{gr} T$ by

$$\operatorname{br} T \leq \operatorname{gr} T. \tag{1}$$

Often, as in the case of the binary tree, equality holds here. However, there are many examples of strict inequality.

Example: If T is a tree such that vertices at even distances from 0 have 2 children while the rest have 3 children, then $\operatorname{br} T = \operatorname{gr} T = \sqrt{6}$.

Example: Let T be a tree embedded in the upper half plane with 0 at the origin. List S_n in clockwise order as $\langle v_1^n, \ldots, v_{2^n}^n \rangle$. Let v_k^n have 1 child if $k \leq 2^{n-1}$ and 3 children otherwise; see Figure 2. Then only one ray (infinite path that doesn't backtrack) consists of vertices having more than one child. This means that water cannot flow down any other ray when $\lambda > 1$, which means it cannot flow down this ray either. Hence $\operatorname{br} T = 1$, yet $\operatorname{gr} T = 2$.

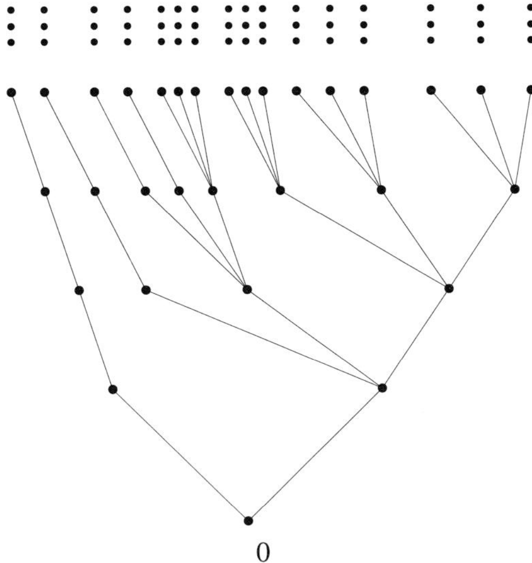

Figure 2.   A tree with branching number 1 and growth rate 2.

Example: If T_1 and T_2 are trees, form a new tree $T_1 \vee T_2$ from disjoint copies of T_1 and T_2 by joining their roots to a new point taken as the root of $T_1 \vee T_2$ (Figure 3). Then

$$\mathrm{br}\,(T_1 \vee T_2) = \mathrm{br}\,T_1 \vee \mathrm{br}\,T_2$$

since water can flow in the join $T_1 \vee T_2$ iff water can flow in one of the trees.

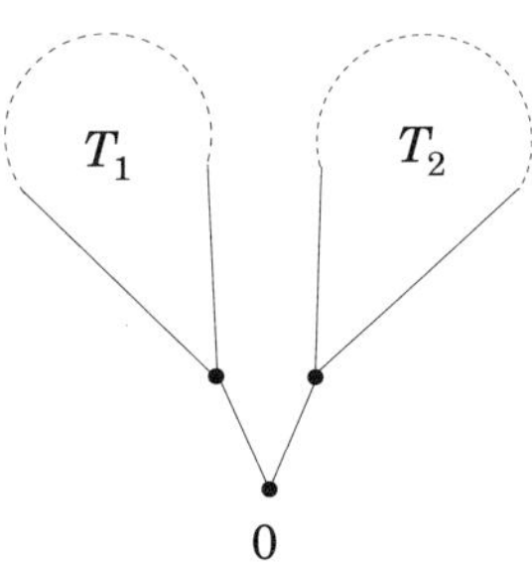

Figure 3.   Joining two trees.

Example: We will put two trees together such that $\mathrm{br}\,(T_1 \vee T_2) = 1$ but $\mathrm{gr}\,(T_1 \vee T_2) > 1$. Let $n_k \uparrow \infty$. Let T_1 (resp., T_2) be a tree such that v has 1 child (resp., 2 children) for $n_{2k} \leq |v| \leq n_{2k+1}$ and 2 (resp., 1) otherwise. If n_k increases sufficiently rapidly, then $\mathrm{br}\,T_1 = \mathrm{br}\,T_2 = 1$, so $\mathrm{br}\,(T_1 \vee T_2) = 1$. But if n_k increases sufficiently rapidly, then $\mathrm{gr}\,(T_1 \vee T_2) = \sqrt{2}$.

While $\mathrm{gr}\,T$ is easy to compute, $\mathrm{br}\,T$ may not be. Fortunately, Furstenberg (1967) gave a useful condition sufficient for equality in (1): Given a vertex v in T,

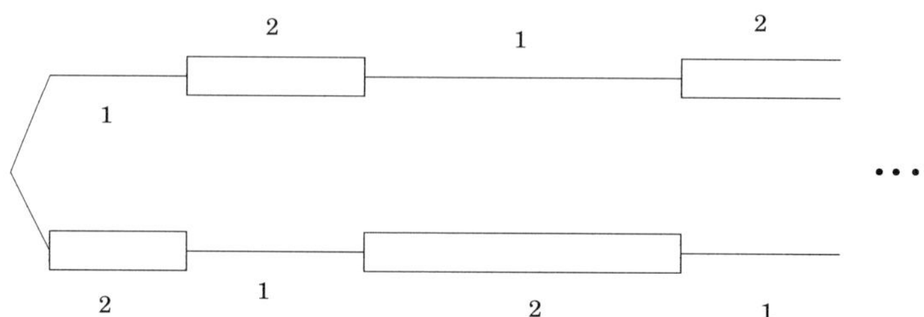

Figure 4. A schematic representation of a tree with branching number 1 and growth
rate $\sqrt{2}$.

let T^v denote the subtree of T formed by the descendants of v. This tree is rooted
at v.

Theorem 1. *If for all vertices $v \in T$, there is an isomorphism of T^v as a rooted
tree to a subtree of T rooted at 0, then* $\operatorname{br} T = \operatorname{gr} T$.

We call trees satisfying the hypothesis of this theorem **subperiodic**. As we
shall see, subperiodic trees arise naturally, which accounts for the importance of
Theorem 1.

2 Electric Current

We can ask another flow question on trees, namely: If λ^{-n} is the conductance of
edges at distance n from the root of T and a battery is connected between the root
and "infinity", will current flow? Of course, what we mean is that we establish a
unit potential between the root and level N of T, let $N \to \infty$, and see whether
the limiting current is positive. If so, the tree is said to have positive effective
conductance and finite effective resistance.

Example: Consider the binary tree. By symmetry, all the vertices at a given distance from 0 have the same potential, so they may be identified without changing
any voltages or currents. This gives a new graph whose vertices may be identified
with $\mathbf{N}$, while there are 2^n edges joining $n - 1$ to n. These edges are in parallel,
so they may be replaced by a single edge whose conductance is their sum, $(2/\lambda)^n$.
Now we have edges in series, so the effective resistance is the sum of the edge
resistances, $\sum_n (\lambda/2)^n$. This is finite iff $\lambda < 2$. Thus, current flows in the infinite
binary tree iff $\lambda < 2$. Note the slight difference to water flow: when $\lambda = 2$, water
can still flow on the binary tree.

In general, there will be a critical value of λ below which current flows and
above which it does not. It turns out that this critical value is the same as that
for water flow (Lyons 1990):

Theorem 2. *If $\lambda < \operatorname{br} T$, then electrical current flows, but if $\lambda > \operatorname{br} T$, then it does
not.*

3   Random Walks

There is a well-known and easily established correspondence between electrical networks and random walks that holds for all graphs. Namely, given a finite connected graph G with conductances assigned to the edges, we consider the random walk which can go from a vertex only to an adjacent vertex and whose transition probabilities from a vertex are proportional to the conductances along the edges to be taken. That is, if v is a vertex with neighbors $u_1, \ldots, u_d$ and the conductance of the edge (v, u_i) is C_i, then $p(v, u_j) = C_j / \sum_{i=1}^{d} C_i$. Now consider two fixed vertices a_0 and a_1 of G. Suppose that a battery is connected across them so that the voltage at a_i equals i $(i = 0, 1)$. Then certain currents will flow along the edges and establish certain voltages at the other vertices in accordance with Kirchhoff's law and Ohm's law. The following proposition provides the basic connection between random walks and electrical networks:

Proposition 1. *For any vertex v, the voltage at v equals the probability that the corresponding random walk visits a_1 before it visits a_0 when it starts at v.*

In fact, the proof of this proposition is simple: there is a discrete Laplacian (a difference operator) for which both the voltage and the probability mentioned are harmonic functions of v. The two functions clearly have the same values at a_i (the boundary points) and the uniqueness principle holds for this Laplacian, whence the functions agree at all vertices v. A superb elementary exposition of this correspondence is given by Doyle and Snell (1984).

What does this say about our trees? Given N, identify all the vertices of level N, i.e., S_N, to one vertex, a_1. Use the root as a_0. Then according to Proposition 1, the voltage at v is the probability that the random walk visits level N before it visits the root when it starts from v. When $N \to \infty$, the limiting voltages are all 0 iff the limiting probabilities are all 0, which is the same thing as saying that on the infinite tree, the probability of visiting the root from any vertex is 1, i.e., the random walk is recurrent. Since no current flows across edges whose endpoints have the same voltage, we see that no electrical current flows iff the random walk is recurrent. Denoting the dependence of the random walk on the parameter λ by RW_λ, we may translate Theorem 2 into the following theorem (Lyons 1990):

Theorem 3. *If $\lambda < \mathrm{br}\, T$, then RW_λ is transient, while if $\lambda > \mathrm{br}\, T$, then RW_λ is recurrent.*

Is this intuitive? Consider the relative weights at a vertex other than the root with, say, d children (see Figure 5). The edge leading back toward the root is λ times as likely to be taken as each other edge. Thus, if we consider only the distance from 0, which increases or decreases at each step of the random walk, a balance between increasing and decreasing occurs when $\lambda = d$. If d were constant, it is easy to see that indeed d would be the critical value separating transience from recurrence. What Theorem 3 says is that this same heuristic can be used, provided we substitute the "average" $\mathrm{br}\, T$ for d.

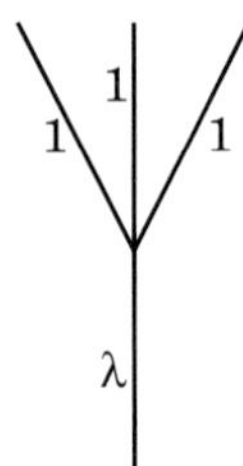

Figure 5. The relative weights at a vertex. The tree is growing upwards.

4   Percolation

Suppose that we remove edges at random from T. To be specific, keep each edge with some fixed probability p and make these decisions independently for different edges. This random process is called **percolation**. By Kolmogorov's 0-1 law, the probability that an infinite connected component remains in the tree is either 0 or 1. On the other hand, this probability is monotonic in p, whence there is a critical value $p_c(T)$ where it changes from 0 to 1. It is also clear that the "bigger" the tree, the more likely it is that there will be an infinite component for a given p. That is, the "bigger" the tree, the smaller the critical value p_c. Thus, p_c is vaguely inversely related to a notion of average branching number. Actually, this vague heuristic is precise (Lyons 1990):

Theorem 4. *For any tree,* $p_c(T) = 1/\mathrm{br}\, T$.

Let us look more closely at the intuition behind this. If a vertex v has d children, then the expected number of children after percolation is dp. If dp is "usually" less than 1, one would not expect that an infinite component would remain, while if dp is "usually" greater than 1, then one might guess that an infinite component would be present. Theorem 4 says that this intuition is precise when one replaces d by $\mathrm{br}\, T$.

Note that there is an infinite component with probability 1 iff the component of the root is infinite with positive probability.

5   Random Trees

Percolation on a fixed tree produces random trees by random pruning, but there is a way to grow trees randomly due to Bienaymé in 1845. Given probabilities p_k adding to 1 ($k = 0, 1, 2, \ldots$), we begin with one individual and let it reproduce according to these probabilities, i.e., it has k children with probability p_k. Each of these children (if there are any) then reproduce independently with the same law, and so on forever or until some generation goes extinct. The family trees produced by such a process are called **(Bienaymé)-Galton-Watson trees**. A fundamental theorem in the subject is that extinction is a.s. iff $m \leq 1$ and $p_1 < 1$, where $m := \sum_k k p_k$ is the mean number of offspring. This provides further justification for the intuition we sketched behind Theorem 4. It also raises a natural question:

Given that a Galton-Watson tree is nonextinct (infinite), what is its branching number? All the intuition suggests that it is m a.s., and indeed it is. This was first proved by Hawkes (1981) and Falconer (1986). But here is the idea of a very simple proof (Lyons 1990).

According to Theorem 4, to determine $\operatorname{br} T$, we may determine $p_c(T)$. Thus, let T grow according to a Galton-Watson process, then perform percolation on T, i.e., keep edges with probability p. We are interested in the component of the root. Looked at as a random tree in itself, this component appears simply as some other Galton-Watson tree; its mean is mp by independence of the growing and the "pruning" (percolation). Hence, the component of the root is infinite w.p.p. iff $mp > 1$. This says that $p_c = 1/m$ a.s. on nonextinction, i.e., $\operatorname{br} T = m$.

6   Branching Random Walk

Given a tree T, assign i.i.d. real-valued displacements to the edges. Let S_v be the sum of the displacements on the path from 0 to v. Denote the distance from v to 0 by $|v|$. We call a path $v_0, v_1, v_2, \ldots$ starting from the root ($v_0 = 0$) a **ray** when v_{i+1} is a child of v_i for every i. Denote by ξ_n the nth point of a ray ξ. Define the **boundary** of the tree, denoted ∂T, as the set of all rays. We may imagine that the rays represent individuals doing random walks on $\mathbb{R}$. Their paths are identical until certain splitting occurs, after which they are independent (and more splitting occurs). This is called either **tree-indexed random walk** or **branching random walk**.

A central question is how fast this branching random walk goes to infinity. Of course, there are many ways to define speed in this context. Some of them are represented by the following quantities:

$$A(T) := \lim_{n \to \infty} \sup_{|v| \geq n} \frac{S_v}{|v|},$$

$$B(T) := \sup_{\xi \in \partial T} \liminf_{n \to \infty} \frac{S_{\xi_n}}{n},$$

$$C(T) := \sup \left\{ \lim_{n \to \infty} \frac{S_{\xi_n}}{n} ;\ \xi \in \partial T,\ \lim_{n \to \infty} \frac{S_{\xi_n}}{n} \text{ exists} \right\}.$$

Then clearly

$$A(T) \geq B(T) \geq C(T).$$

Let X be a random variable with the law of the displacements. It has a "rate function"

$$\mu(s) := \inf_{t \leq 0} \mathbf{E}\left[e^{t(s-X)} \right],$$

with an "inverse"

$$\mu_1(a) := \sup\{ s ;\ \mu(s) < a \}.$$

A special case of a fundamental theorem of Biggins (1976), often rediscovered, identifies the speed for Galton-Watson trees:

Theorem 5. *If T is a Galton-Watson tree with mean m, then*

$$A(T) = B(T) = C(T) = \mu_1(1/m)$$

a.s. given nonextinction.

This can be partially generalized to any tree (Lyons and Pemantle 1992):

$$B(T) = C(T) = \mu_1(1/\operatorname{br} T) \quad \text{a.s.}$$

It is not generally true that $A(T)$ has the same value, however.

7 Hausdorff Dimension

Consider again any finite connected graph with two distinguished vertices a_0 and a_1. This time, the edges e have assigned positive numbers $C(e)$ that represent the maximum amount of water that can flow through the edge (in either direction). How much water can flow into a_0 and out of a_1? Consider any set W of edges that separates a_0 from a_1, i.e., the removal of all edges in W would leave a_0 and a_1 in different components. Such a set W is called a **cutset**. Since all the water must flow through W, an upper bound for the maximum flow from a_0 to a_1 is $\sum_{e \in W} C(e)$. The beautiful Max-flow Min-cut Theorem of Ford and Fulkerson (1962) says that these are the only constraints: the maximum flow equals $\inf_{W \text{ a cutset}} \sum_{e \in W} C(e)$.

Applying this theorem to our tree situation, we see that the maximum flow from the root to infinity is

$$\inf \left\{ \sum_{v \in \Pi} \lambda^{-|v|} \, ; \, \Pi \text{ cuts the root from infinity} \right\}.$$

Here, we identify a set of vertices Π with their preceding edges when considering cutsets.

Now there is a natural metric on ∂T: if $\xi, \eta \in \partial T$ have exactly n edges in common, define their distance to be $d(\xi, \eta) := e^{-n}$. Thus, if $v \in T$ has more than one child with infinitely many descendants, the set of rays going through v,

$$B_v := \left\{ \xi \in \partial T \, ; \, \xi_{|v|} = v \right\},$$

has diameter $|B_v| = e^{-|v|}$. We call a collection $\mathcal{C}$ of subsets of ∂T a **cover** if

$$\bigcup_{B \in \mathcal{C}} B = \partial T.$$

Note that Π is a cutset iff $\{B_v \, ; \, v \in \Pi\}$ is a cover. The **Hausdorff dimension** of ∂T is defined to be

$$\dim T := \sup \left\{ \alpha \, ; \, \inf_{\mathcal{C} \text{ a cover}} \sum_{B \in \mathcal{C}} |B|^\alpha > 0 \right\}.$$

This is, in fact, already familiar to us, since

$$\operatorname{br} T = \sup\{\lambda\,;\ \text{water can flow through pipe capacities } \lambda^{-|v|}\}$$

$$= \sup\left\{\lambda\,;\ \inf_{\Pi\ \text{a cutset}} \sum_{v\in\Pi} \lambda^{-|v|} > 0\right\}$$

$$= \exp\sup\left\{\alpha\,;\ \inf_{\Pi\ \text{a cutset}} \sum_{v\in\Pi} e^{-\alpha|v|} > 0\right\}$$

$$= \exp\sup\left\{\alpha\,;\ \inf_{C\ \text{a cover}} \sum_{B\in C} |B|^{\alpha} > 0\right\}$$

$$= \exp\dim\partial T\,.$$

8 Capacity

An important tool in analyzing electrical networks is that of energy. Thompson's principle says that, given a finite graph and two distinguished vertices a_0, a_1, the unit current flow from a_0 to a_1 is the unit flow from a_0 to a_1 that minimizes energy, where, if the conductances are $C(e)$, the **energy** of a flow θ is defined to be $\sum_{e\ \text{an edge}} \theta(e)^2/C(e)$. (It turns out that the energy of the unit current flow is equal to the effective resistance from a_0 to a_1.) Thus,

$$
\begin{array}{c}
\text{electrical current flows from the root of an infinite tree} \\
\Longleftrightarrow \\
\text{there is a flow with finite energy.}
\end{array}
\tag{2}
$$

Now on a tree, a unit flow can be identified with a function θ on the vertices of T that is 1 at the root and has the property that for all vertices v,

$$\theta(v) = \sum_i \theta(u_i)\,,$$

where u_i are the children of v. The energy of a flow is then

$$\sum_{v\in T} \theta(v)^2 \lambda^{|v|}\,,$$

whence

$$\operatorname{br} T = \sup\left\{\lambda\,;\ \text{there exists a unit flow } \theta\ \sum_{v\in T} \theta(v)^2 \lambda^{|v|} < \infty\right\}\,.\tag{3}$$

We can also identify unit flows on T with Borel probability measures μ on ∂T via

$$\mu(B_v) = \theta(v)\,.$$

A bit of algebra shows that (3) is equivalent to

$$\operatorname{br} T = \exp\sup\left\{\alpha\,;\, \exists \text{ a probability measure } \mu \text{ on } \partial T \quad \iint \frac{d\mu(\xi)\,d\mu(\eta)}{d(\xi,\eta)^\alpha} < \infty\right\}.$$

Define **capacity** to be the reciprocal of the minimum energy:

$$\operatorname{cap}_\alpha(\partial T)^{-1} := \inf\left\{\iint \frac{d\mu(\xi)\,d\mu(\eta)}{d(\xi,\eta)^\alpha}\,;\, \mu \text{ a probability measure on } \partial T\right\}.$$

Then statement (2) says that for $\alpha > 0$,

$$\text{random walk with parameter } \lambda = e^\alpha \text{ is transient} \iff \operatorname{cap}_\alpha(\partial T) > 0\,. \qquad (4)$$

It follows from Theorem 3 that

$$\text{the critical value of } \alpha \text{ for positivity of } \operatorname{cap}_\alpha(\partial T) \text{ is } \dim \partial T\,. \qquad (5)$$

A refinement of Theorem 4 is (Lyons 1992):

Theorem 6. *Percolation with parameter $p = e^{-\alpha}$ yields an infinite component a.s. iff $\operatorname{cap}_\alpha(\partial T) > 0$. Moreover,*

$$\operatorname{cap}_\alpha(\partial T) \leq \mathbf{P}[\text{the component of the root is infinite}] \leq 2\operatorname{cap}_\alpha(\partial T)\,.$$

The case of the first part of this theorem where all the degrees are uniformly bounded was shown first by Fan (1989, 1990).

One way to use this theorem is to combine it with (4); this allows us to translate problems freely between the domains of random walks and percolation (Lyons 1992). The theorem actually holds in a much wider context (on trees) than that mentioned here: Fan allowed the probabilities p to vary depending on the generation and Lyons allowed the probabilities as well as the tree to be completely arbitrary.

9 Embedding Trees Into Euclidean Space

The results described above, especially those concerning percolation, can be translated to give results on closed sets in Euclidean space. We shall describe only the simplest such correspondence here, which was the one that was part of Furstenberg's motivation in 1970. Namely, for a closed nonempty set $E \subseteq [0,1]$ and for any integer $b \geq 2$, consider the system of b-adic subintervals of $[0,1]$. Those whose intersection with E is non-empty will form the vertices of the associated tree. Two such intervals are connected by an edge iff one contains the other and the ratio of their lengths is b. The root of this tree is $[0,1]$. We denote the tree by $T_b(E)$. Were it not for the fact that certain numbers have two representations in base b, we

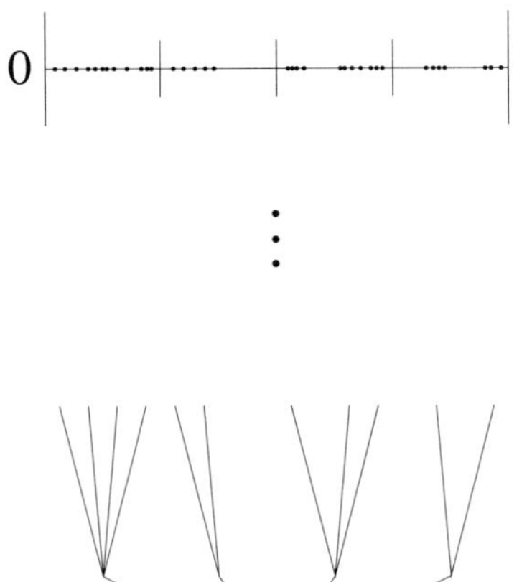

Figure 6.   In this case $b = 4$.

could identify $\partial T_b(E)$ with E. Note that such an identification would be Hölder continuous in one direction (only).

Hausdorff dimension is defined for subsets of $[0, 1]$ just as it was for ∂T:

$$\dim E := \sup \left\{ \alpha \,;\; \inf_{\mathcal{C} \text{ a cover of } E} \sum_{B \in \mathcal{C}} |B|^\alpha > 0 \right\},$$

where $|B|$ denotes the (Euclidean) diameter of E. Covers of $\partial T_b(E)$ by sets of the form B_v correspond to covers of E by b-adic intervals. It easy to show that restricting to such covers does not change the computation of Hausdorff dimension, whence we may conclude that

$$\dim E = \frac{\dim \partial T_b(E)}{\log b} = \log_b(\operatorname{br} T_b(E)).$$

Capacity is also defined as it was on the boundary of a tree:

$$(\operatorname{cap}_\alpha E)^{-1} := \inf \left\{ \iint \frac{d\mu(x)\, d\mu(y)}{|x - y|^\alpha} \,;\; \mu \text{ a probability measure on } E \right\}.$$

It was shown by Benjamini and Peres (1992) that

$$(\operatorname{cap}_\alpha E)/3 \le \operatorname{cap}_{\alpha \log b}(\partial T_b(E)) \le b \operatorname{cap}_\alpha E. \tag{6}$$

This means that the percolation criterion Theorem 6 can be used in Euclidean space. In fact, inequalities such as (6) also hold for more general kernels than distance to a power and for higher-dimensional Euclidean spaces. Nice applications to the path of Brownian motion were found by Peres (1996) by replacing the path by an "intersection-equivalent" random fractal that is much easier to analyze, being an embedding of a Galton-Watson tree.

Also, the fact (5) translates into the analogous statement about subsets $E \subseteq [0, 1]$; this is a classical theorem of Frostman (1935).

10 Cayley Graphs

Suppose we investigate RW_λ on graphs other than trees. What do we mean? Fix a vertex 0 in a graph G. We always assume that G is connected and that there are only a finite number of edges incident to each vertex. If e is an edge whose endpoint furthest from 0 is at distance n, let the conductance of e be λ^{-n}. The argument we used for trees in Section 3 to see that transience is equivalent to positive current flow also applies to arbitrary graphs. In order to understand what the critical value $\lambda_c(G)$ separating transience from recurrence measures, consider the class of spherically symmetric graphs, where we call G **spherically symmetric** about 0 if for all pairs of vertices v, w at the same distance from 0, there is an isomorphism of G fixing 0 that takes v to w. Let $\tilde{M}_n$ be the number of edges that lead from a vertex at distance $n-1$ from 0 to a vertex at distance n. Then the critical value of λ is the growth rate of G:

$$\lambda_c(G) = \liminf_{n\to\infty} \tilde{M}_n^{1/n}\,.$$

In fact, we have the following more precise criterion for transience:

Proposition 2. *If G is spherically symmetric about 0, then RW_λ is transient iff* $\sum_n \lambda^n/\tilde{M}_n < \infty$.

Proof. Since G is spherically symmetric, all vertices at a given distance from 0 have the same voltage. Therefore, they can be identified without changing any voltages (or currents). This yields the graph $\mathbf{N}$ with multiple edges and loops. We may ignore the loops; and there are $\tilde{M}_n$ edges between $n-1$ and n. Since they are in parallel, they are equivalent to a single edge whose conductance is the sum of the conductances, $\tilde{M}_n\lambda^{-n}$. These new edges are in series, so they are equivalent to a single edge (from 0 to infinity) with resistance the sum of the resistances, $\sum_n \lambda^n/\tilde{M}_n$. This gives the proposition. $\square$

Next, consider the **Cayley graphs** of finitely generated groups: if G is a group generated by $\{g_1, \ldots, g_k\}$ as a semigroup, then we form a graph with vertices G and edges $\{(v, vg_i)\,;\ v \in G,\ 1 \le i \le k\}$. Because $\{g_i\}$ generates G, the graph is connected. It is also **homogeneous**, that is, its automorphism group is transitive on vertices. Some examples of Cayley graphs with natural generating sets appear in Figure 7.

Although Cayley graphs are homogeneous, they are generally not spherically symmetric. Still, if M_n denotes the number of vertices at distance n from the identity, then $\lim M_n^{1/n}$ exists since the triangle inequality implies that $M_{m+n} \le M_m M_n$. Is $\lambda_c(G) = \lim M_n^{1/n}$?

If we identify the vertices at a given distance from the identity, as in the proof of Proposition 2, we may not preserve the effective conductance from the identity to infinity, but we cannot decrease it. Hence, the calculation in the proof of Proposition 2 shows that if $\sum_n \lambda^n/M_n = \infty$, then RW_λ is recurrent. In particular, this is the case when $\lambda > \lim M_n^{1/n}$. What about when $\lambda < \lim M_n^{1/n}$? It turns out that RW_λ is then transient.

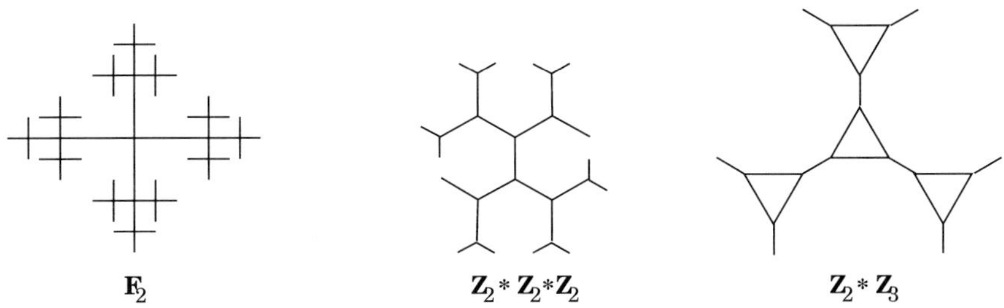

Figure 7.   The Cayley graphs of the free group on 2 letters; the free product of $\mathbb{Z}_2$ with itself 3 times $\mathbb{Z}_2 * \mathbb{Z}_2 * \mathbb{Z}_2$; and the free product $\mathbb{Z}_2 * \mathbb{Z}_3$.

To prove transience for a given λ, it suffices to prove that a subgraph is transient: this is the same as saying that removing edges cannot increase effective conductance. The easiest subgraph to analyze would be a subtree of G, while in order to have the greatest likelihood of being transient, it should be as big as possible, i.e., a **spanning tree** (one that includes every vertex). Here is one: for each $v \in G$, there is a unique word $(g_{i_1}, g_{i_2}, \ldots, g_{i_n})$ such that $v = g_{i_1} g_{i_2} \cdots g_{i_n}$, $n = |v|$, and $(g_{i_1}, \ldots, g_{i_n})$ is **lexicographically minimal** with these properties, i.e., if $(g_{i'_1}, \ldots, g_{i'_n})$ is another and m is the first j such that $i_j \neq i'_j$, then $i_m < i'_m$. Call this word w_v. Let T be the subgraph of G containing all vertices and with u adjacent to v when $|u| + 1 = |v|$ and w_u is an initial segment of w_v (or vice versa). We root T at the identity.

Suppose that w_v is broken arbitrarily as the concatenation of two words w_1 and w_2. Let v_i be the product of the generators in w_i $(i = 1, 2)$. Then $v = v_1 v_2$. If for some i, we had $w_{v_i} \neq w_i$, then we could substitute the word w_{v_i} for w_i and find another word w whose product was v yet would be either shorter than w_v or come earlier lexicographically than w_v. Either of these circumstances would contradict the definition of w_v as minimal, whence $w_i = w_{v_i}$ for both i.

This fact for $i = 1$ shows that T is a tree; for $i = 2$, it shows that T is subperiodic. Since T is spanning and since distances to the identity in T are the same as in G, we have $\mathrm{gr}\, T = \lim M_n^{1/n}$. Since T is subperiodic, Furstenberg's Theorem 1 shows that $\mathrm{br}\, T = \lim M_n^{1/n}$. Hence RW_λ is transient on T for $\lambda < \lim M_n^{1/n}$ by Theorem 3, whence on G as well. This gives the following theorem of Lyons (1995):

Theorem 7. RW_λ *on a Cayley graph has critical value* λ_c *equal to the growth rate of the graph.*

Of course, this theorem makes Cayley graphs look spherically symmetric from a probabilistic point of view. Such a conclusion, however, would be too hasty, for there are Cayley graphs with the following very surprising property (Lyons, Pemantle and Peres 1996): define the **speed** (or rate of escape) of RW_λ as the limit of the distance from the identity at time n divided by n as $n \to \infty$, if the limit exists. On any spherically symmetric graph, the speed of RW_λ (if it exists) is monotone decreasing in λ. However, there is a Cayley graph such that the speed is 0 when $\lambda = 1$ but is strictly positive when $1 < \lambda < (1 + \sqrt{5})/2$.

References

BENJAMINI, I. and PERES, Y. (1992) Random walks on a tree and capacity in the interval, *Ann. Inst. Henri Poincaré Probab. Statist.* **28**, 557–592.

BIGGINS, J. D. (1976) The first- and last-birth problems for a multitype age-dependent branching process, *Adv. Appl. Prob.* **8**, 446–459.

DOYLE, P. and SNELL, J. L. (1984) *Random Walks and Electric Networks.* Mathematical Assoc. of America, Washington, D. C.

FALCONER, K. J. (1986) Random fractals, *Math. Proc. Cambridge Philos. Soc.* **100**, 559–582.

FAN, A. H. (1989) *Décompositions de Mesures et Recouvrements Aléatoires.* Thesis, Université de Paris-Sud, Orsay, France.

FAN, A. H. (1990) Sur quelques processus de naissance et de mort, *C. R. Acad. Sci. Paris* **310**, I, 441–444.

FORD, L. R. JR. and FULKERSON, D. R. (1962) *Flows in Networks.* Princeton University Press, Princeton, NJ.

FROSTMAN, O. (1935) Potentiel d'équilibre et capacité des ensembles avec quelques applications à la théorie des fonctions, *Meddel. Lunds Univ. Mat. Sem.* **3**, 1–118.

FURSTENBERG, H. (1967) Disjointness in ergodic theory, minimal sets, and a problem in diophantine approximation, *Math. Systems Theory* **1**, 1–49.

FURSTENBERG, H. (1970) Intersections of Cantor sets and transversality of semigroups, *Problems in analysis (Sympos. Salomon Bochner, Princeton Univ., 1969)*, pp. 41–59. Princeton Univ. Press, Princeton, NJ.

HAWKES, J. (1981) Trees generated by a simple branching process, *J. London Math. Soc.* **24**, 373–384.

LYONS, R. (1990) Random walks and percolation on trees, *Ann. Probab.* **18**, 931–958.

LYONS, R. (1992) Random walks, capacity, and percolation on trees, *Ann. Probab.* **20**, 2043–2088.

LYONS, R. (1995) Random walks and the growth of groups, *C. R. Acad. Sci. Paris*, **320**, 1361-1366.

LYONS, R. and PEMANTLE, R. (1992) Random walk in a random environment and first-passage percolation on trees, *Ann. Probab.* **20**, 125–136.

LYONS, R., PEMANTLE, R., and PERES, Y. (1996) Random walks on the lamplighter group, *Ann. Probab.*, to appear.

PERES, Y. (1996) Intersection-equivalence of Brownian paths and certain branching processes, *Commun. Math. Phys.*, **177**, 417–434.

Russell Lyons
Department of Mathematics
Indiana University
Bloomington, IN 47405-5701
USA

Progress in Probability, Vol. 40

Functional Limit Theorems for the Simple Random Walk on a Supercritical Galton-Watson Tree

DIDIER PIAU

Mathematics Subject Classification: Primary: 60J80. Secondary: 60J15

Keywords: Trees, Galton-Watson, branching processes, random walk in random environment, rate of escape

Abstract. We prove a functional central limit theorem for the range and speed of the simple random walk on the family tree of a Galton-Watson process when the vertex progeny $Z \geq 2$. We reduce the general case $Z \geq 0$ of a Galton-Watson tree conditioned on its non-extinction to the case $Z \geq 1$.

Résumé. Nous démontrons un théorème central limite fonctionnel pour le rang et la vitesse de la marche au hasard simple sur l'arbre généalogique d'un processus de Galton-Watson quand le nombre de descendants est donné par $Z \geq 2$. Nous réduisons le cas général $Z \geq 0$ d'un arbre de Galton-Watson conditionné par sa non extinction au cas $Z \geq 1$.

1   Introduction, statement of results

Let $Z \geq 0$ be an integer valued random variable with finite mean $E(Z) > 1$. The Galton-Watson law G is the usual probability measure on the set of rooted trees t, describing the genealogical (or historical) process starting from a single ancestor $\mathrm{root}(t)$ when the progeny law is given by Z. The (random) tree t is constructed as follows: to the root $\mathrm{root}(t)$, we associate Z vertices, to each of these, we independently associate Z vertices, and so on, all random variables being independent copies of Z. Finally, an edge is drawn between any vertex and any of its children.

Now, for each rooted tree t, one can perform a simple random walk $(x_n)_n$ on the tree t: $(x_n)_n$ is the Markov chain on the vertices of t starting from $x_0 = \mathrm{root}(t)$ and whose probability transitions are

$$p(x, y) = 1/\deg(x)$$

if y is a neighbour of the vertex x, and $p(x, y) = 0$ otherwise. What can be said of the asymptotic behavior of $(x_n)_n$ when one chooses t at random with the Galton-Watson law?

If $P(Z = 0) > 0$ and one conditions on the extinction of the family process, then t is finite and $(x_n)_n$ is recurrent, just like the simple random walk on any finite connected graph. The asymptotic properties of such a Markov chain are relatively well known: up to time n, x_n spends a time proportional to $n \cdot \deg(x)$ at each vertex x. We study the other case, that is we suppose: either (i) $P(Z = 0) > 0$ and one conditions on the non-extinction of the family process; or (ii) $Z \geq 1$ almost surely. The tree t is then almost surely infinite and $(x_n)_n$ is known to be transient for almost every tree t. We further study the random walk $(x_n)_n$ through its range and speed. At time n, the range r_n is the cardinal of the set $\{x_k \, ; \, 1 \le k \le n\}$ while the speed is described using the distance to a given vertex, for example the graph distance d_n between the positions x_n and x_0 of the random walk at times n and 0. The idea is that r_n and d_n provide a sharper description of how the random walk escapes from $\mathrm{root}(t)$. Let us now state the properties of r_n and d_n that we shall prove.

Denote by P_t the law of the random walk $(x_n)_n$ on a given rooted tree t and by GP the probability measure describing the successive picking of t at random with the law G and picking a path $(x_n)_n$ on t at random with the law P_t. The construction of GP is due to R. Lyons, R. Pemantle and Y. Peres [LPP95] and we refer to this paper for more details on GP. A strong law of large numbers for d_n is proved in [LPP95] and we noticed in [Pi94] that the same method also yields a strong law of large numbers for r_n. Hence d_n/n (resp. r_n/n) converges GP-almost surely to a strictly positive constant l (resp. m) given by:

$$l = E((Z - 1)/(Z + 1)), \qquad m = G(C/(1 + C)) = G(C)/E(Z),$$

where C is the effective conductance of the tree t between its root and its boundary (for the notion of effective conductance of a graph, see [DS84]).

The aim of this paper is to prove a functional central limit theorem for r_n and d_n. Denote by $r^{(n)}$ the interpolation of the range at time $n \ge 1$; this is a random process defined for $0 \le s \le 1$ by the following formula:

$$r^{(n)}(s) = n^{-1/2} \left(r_{[ns]} - m \cdot ns \right),$$

where $[ns]$ is the integer part of ns. With a similar definition for $d^{(n)}$, we prove:

Theorem A – *Assume that $Z \ge 2$ almost surely. Then, the range and the speed of simple random walk on the genealogical Galton-Watson tree with progeny law Z converge in law towards a Brownian motion in the following sense: there exist two real numbers $\lambda > 0$ and $\mu > 0$ such that $\mu^{-1} r^{(n)}(\cdot)$ and $\lambda^{-1} d^{(n)}(\cdot)$ converge in law, under GP and when n goes to infinity, towards $(W_t)_{0 \le t \le 1}$ where W is a linear Brownian motion starting at zero. The convergence is in the sense of uniform topology on the Skorokhod space of right continuous functions with left limits.*

We don't know any expression for the values of m, λ and μ, in terms of Z alone. However, when Z is constant, $Z = p$ with $p \ge 2$, the tree is regular of degree $p+1$; we get easily $l = (p-1)/(p+1)$, $m = (p-1)/p$. It is also possible to compute λ and μ in this degenerate case (see section 3.1):

$$\lambda^2 = 1 - l^2, \quad \mu^2 = m(1 - m) + 2 \sum_{k \ge 1} (p^k (1 + p^k))^{-1}.$$

As for the general case, we don't know how to prove theorem A neither in case (i) $P(Z = 0) > 0$ and we condition on non-extinction, nor in case (ii) $Z \geq 1$ almost surely and $P(Z = 1) > 0$. We formulate conjecture B in order to state a partial result in this direction (theorem C).

Conjecture B – *Suppose $Z \geq 1$ almost surely with $P(Z = 1) > 0$ and let n_R be the first regeneration time (see section 3.1). There exists constants $a > 0$ and $c > 0$ such that, for any n,*

$$GP(n_R \geq n) \leq \exp(-c\,n^a).$$

When $Z \geq 2$, the conclusion of conjecture B is true for $a = 1/2$ and this fact implies theorem A. When $P(Z = 1) > 0$, it is false for any $a > 1/3$ (Y. Peres, private communication). However theorem C shows that the case (ii) is the only one we need to solve. We skip the proof of this result.

Theorem C – *Suppose $P(Z = 0) > 0$. There exists a random variable $Z' \geq 1$ with finite mean $E(Z') = E(Z) > 1$ such that conjecture B is true for GP as soon as it is for the law $G'P$ associated to Z'.*

Hence, the range and distance of the random walk satisfy a functional central limit theorem under GP as soon as conjecture B is true for the law $G'P$.

Theorem A is true for every law GP if conjecture B is true for the law GP associated to any $Z \geq 1$. Let us now describe the typical situation one has to solve in order to prove conjecture B. Assume that the law of Z is

$$p\delta_1 + (1-p)\delta_2, \qquad p \in (0,1).$$

Then we construct the tree t under G as follows: consider the binary tree b where every vertex has degree 3, replace each edge e of b by a linear portion of $l(e)$ successive edges, $l(e) \geq 1$, and assume that the lengths $l(e)$ are i.i.d. and that the law of $l(e)$ is geometric:

$$P(l(e) = n) = (1-p)\,p^{n-1}, \qquad n \geq 1.$$

We claim that the law of the resulting tree is GP and we guess that a proof of conjecture B in this particular case yields a proof of the general case.

The paper is organized as follows. In section 2, we recall the construction of the law GP which ensures that the range and distance satisfy a strong law of large numbers. This part relies heavily on [LPP95]. In section 3, we prove theorem A. Some of the results of this paper were announced in [Pi95].

We wish to thank gratefully R. Lyons, R. Pemantle and Y. Peres for sending us [LPP95] while their paper was still a preprint and for instructive discussions, and A. Rouault for inviting us to lecture on these results at "Trees '95".

2   Paths-on-trees, strong laws

We recall the construction of simple random walk on a (random) Galton-Watson tree due to R. Lyons, R. Pemantle and Y. Peres, see [LPP95].

Let $Z \geq 0$ be an integer valued random variable with finite mean $E(Z) > 1$ and $p_n = P(Z = n)$ for any $n \geq 0$. T is the set of infinite rooted trees t, that is of locally finite connected graphs without cycles where a given vertex $\mathrm{root}(t)$ is chosen and called the root of t. G is the law on T of the family tree t of the branching process associated to the law of Z starting from a single ancestor $\mathrm{root}(t)$. When $p_0 = P(Z = 0) > 0$, the branching process has a probability $q \in (0,1)$ to become extinct, so that in this case G is the law of the tree associated to the branching process conditioned by its non-extinction. Note that $E(Z) > 1$, hence the process is supercritical and one conditions by an event of strictly positive probability.

For a given tree t, the simple random walk $(x_n)_n$ on t is the Markov chain on the set of the vertices of t starting from $x_0 = \mathrm{root}(t)$ such that x_{n+1} is one of the neighbours of x_n, equiprobably chosen. We modify the usual definition of G in the following way: each vertex of t is given Z children in law, except $\mathrm{root}(t)$ which is given $Z + 1$ children in law. Without this modification, the property of stationarity of G which we will use is false because of the singular status of the root. Nevertheless, because of the transience, all the asymptotic results concerning the random walk on t are true for this "augmented" Galton-Watson law as well as for the "non-augmented" one. From now on, G denotes the "augmented" Galton-Watson law.

The measure GP describing the paths of simple random walk $(x_n)_n$ on a random tree t chosen by the law G *is not* a product measure but GP can be constructed as follows. Let B be the bundle over T made of the $(t, \hat{x})$ with $t \in T$ and $\hat{x} = (x_n)_n$ any bi-infinite sequence of vertices of t such that $x_0 = \mathrm{root}(t)$ and that x_n and x_{n+1} are neighbours. Denote by

$$\pi(t, t') = 1/\deg(\mathrm{root}(t))$$

if $t, t' \in T$ and if t' has the same underlying graph than t but with $\mathrm{root}(t')$ located at one of the neighbours of $\mathrm{root}(t)$, and $\pi(t, t') = 0$ otherwise. Then π is the transition kernel of a Markov chain $(t_n)_n$ on T and G is stationary reversible for π, see [LPP95, theorem 3.1]. Let GP be the measure on B associated to π and G. The path $(x_n)_n$ is simply the sequence of the roots $(\mathrm{root}(t_n))_n$. A shift map on B is defined as

$$S(t, \hat{x}) = (u, \hat{y}),$$

if $y_n = x_{n+1}$ for any n and if u is the tree t with its root located at x_1. The key observation of [LPP95] is that S is ergodic under GP. Birkhoff ergodic theorem may then be applied to the sum of the iterates of a given random variable on B under the action of S. The distance d_n, see [LPP95], and the range r_n, see [Pi94], are almost exactly the values of such sums, hence they satisfy strong laws of large numbers.

3   Proof of theorem A

We first prove a simple central limit theorem (convergence to a centered normal law) by using the subset R of B made of the so-called regeneration points. For this method to be applicable, the return time n_R to R must be square integrable. We show that the tail of the law of n_R decreases subexponentially when $Z \geq 2$ almost surely. This bound allows us to extend the convergence to a normal law to a functional central limit theorem, which is the convergence to a Brownian motion. This extension uses a result of D. Scott, see [Sco73].

We give the proof for the range and mention the minor modifications that allow to treat the case of the distance too. We write $\mathbf{1}_B$ or $\mathbf{1}(B)$ for the indicator function of a set B; $P(X)$ for the expectation of a random variable X under a probability measure P; and $a(n)$ instead of a_n for all kinds of symbols a and n.

3.1 Convergence to a normal law

In section 2, a shift operator S on B was defined, which shifts the time of $+1$ by sending $(t, \hat{x})$ to $(St, S\hat{x})$ as follows: St is the rooted tree with the same underlying graph than t and whose root is located at x_1 and $(S\hat{x})_n = x_{n+1}$ for all n. The measure GP is invariant under the action of S as well as under the reversal of time

$$(t, (x_n)_n) \mapsto (t, (x_{-n})_n).$$

Let R be the subset of B defined by

$$R = \{\forall n \geq 0,\, \forall m \leq -1,\, x_n \neq x_{-1},\, x_m \neq x_0\}.$$

A path-on-a-tree $(t, \hat{x})$ is in R if and only if the random walk just crossed the edge (x_{-1}, x_0) for the first and last time. The measure of R is strictly positive, hence

$$n_R(t, \hat{x}) = \inf\{n \geq 1\,;\, S^n(t, \hat{x}) \in R\}$$

is finite for GP almost every $(t, \hat{x})$ as well as the successive passage times n_R^k defined by $n_R^1 = n_R$ and

$$n_R^{k+1} = n_R \circ S^{n_R^k}.$$

Call the portion of the path between the times n_R^k and n_R^{k+1}, the k-th slab. The slabs are i.i.d. for any $k \geq 1$. Consider $\bar{r}_k$ the range of the k-th slab and $(n_k)_k$ the counting process associated to the sequence $(n_R^k)_k$. Precise definitions are as follows:

$$\bar{r}_k = \#\{x_n\,;\, n_R^k \leq n < n_R^{k+1}\}, \qquad \{n_k \geq p\} = \{k \geq n_R^p\}.$$

The range at time $k \geq 1$ is

$$r_k = a_k + \sum_{l=1}^{n_k - 1} \overline{r}_l + b_k$$

with $a_k = r(k \wedge n_R)$ and $b_k = (r_k - r(n_R^{n_k}))\, \mathbf{1}_{n \geq n_R}$. The range of a path during a given time interval is at most the length of this time interval, hence $a_k \leq n_R$ and $a_k / \sqrt{k}$ converges almost surely to 0. For the same reason, $b_k \leq \overline{r}_{n_k}$: this is the length of the overlapping interval in the renewal process associated to a sequence $(\overline{r}_k)_k$ and to a first delaying term $r(n_R)$. Of course, the law of this overlapping length is not the common law of the $\overline{r}_k$'s but it converges in law as k goes to infinity to a random variable $\overline{r}_\infty$, see [Dur91, section 3.4] for example. The law of $\overline{r}_\infty$ can be characterized by

$$GP(f(\overline{r}_\infty)) = GP(\overline{r}_1\, f(\overline{r}_1))/GP(\overline{r}_1)$$

for any measurable nonnegative f. Hence $b_k / \sqrt{k}$ converges in law to 0 as soon as $\overline{r}_\infty$ is almost surely finite, and this is obviously true because $\overline{r}_1$ is integrable.

We now look at the sum of the ranges of the slabs. By Birkhoff ergodic theorem applied to S^{n_R} (or by the strong law of large numbers applied to the $\overline{r}_k$'s), n_k / k converges GP almost surely to a strictly positive constant ν. Let c_k be the sum

$$c_k = \sum_{l=1}^{n_k - 1} \overline{r}_l - m\,(n_R^{l+1} - n_R^l) = \sum_{l=1}^{n_k - 1} \widetilde{r}_l.$$

If the $\widetilde{r}_k$'s are square integrable, $c_k / \sqrt{k}$ converges in law to a centered normal law. Now the centered range $\widetilde{r}_k$ is bounded by $(n_R^{k+1} - n_R^k)$, which follows the law of n_R conditioned by R. Hence the square integrability of n_R proves the convergence to a centered normal law for the range.

The proof for the distance is analogous: a tree has no cycles hence the shortest path from x_0 to x_k goes through all the regeneration edges through which the random walk has already gone. Define the distance of a slab as the graph distance $\overline{d}_k$ between $x(n_R^{k+1})$ and $x(n_R^k)$, and the centered distance of a slab as

$$\widetilde{d}_k = \overline{d}_k - l\,(n_R^{k+1} - n_R^k).$$

Then $\widetilde{d}_k$ is bounded by $(n_R^{k+1} - n_R^k)$ hence the same condition on n_R ensures the convergence to a centered normal law for the distance as well.

3.2 Values of the variances

We assume that n_R is integrable and give an expression of the parameters λ and μ. The error terms a_k and b_k can be neglected and n_k is equivalent to $k\nu$ for

$$\nu = GP(n_R \,|\, R) = GP(R)^{-1} = m^{-2}.$$

Hence λ and μ are given by

$$\lambda^2 = \nu \operatorname{var}(\widetilde{d}_1), \qquad \mu^2 = \nu \operatorname{var}(\widetilde{r}_1).$$

The random variables $\widetilde{d}_1$ and $\widetilde{r}_1$ are never degenerate, because this would imply for instance that the range of a slab is proportional to its time-length. Hence λ^2 and μ^2 are strictly positive.

To get the values of λ^2 and μ^2 explicitly from the law of Z is an open question in the general case. When Z is almost surely constant, $P(Z = p) = 1$ for a $p \geq 2$, the tree is deterministic and regular, hence one can apply the techniques of a paper by Jain and Pruitt [JP71]. These authors study the range of simple random walk on the d dimensional square lattice, $d \geq 3$. The two essential properties they use are the absolute invariance of the environment around the position of the random walk and the symmetry of the transition probabilities. These properties are true in our setting so that, following their method, one can compute the limit of the variance of $r_k/\sqrt{k}$, getting

$$l = (p-1)/(p+1), \qquad m = (p-1)/p,$$

$$\mu^2 = m(1-m) + 2\sum_{k\geq 1}(p^k(1+p^k))^{-1}.$$

We skip these computations because they follow closely [JP71] and depend basically on easy computations for certain random walks on the line. We now compute λ^2 more directly. Let $q_n(k)$ be the probability for the distance between x_n and x_0 to be k. One has:

$$q_{n+1}(k) = \begin{cases} \frac{1}{2}(1+l)q_n(k-1) + \frac{1}{2}(1-l)q_n(k+1) & \text{for } k \geq 2, \\ q_n(0) + \frac{1}{2}(1-l)q_n(2) & \text{for } k = 1, \\ \frac{1}{2}(1-l)q_n(1) & \text{for } k = 0. \end{cases}$$

It follows that

$$E(d_{n+1}) = E(d_n) + l + \varepsilon_n,$$

$$\operatorname{var}(d_{n+1}) = \operatorname{var}(d_n) + 1 - l^2 + \eta_n,$$

for $\varepsilon_n = (1-l)\, q_n(0)$ and $\eta_n = 2\varepsilon_n\,(l + E(d_n)) + \varepsilon_n^2$. Direct computations (or lemma 1 of section 3.3, which is easy for a regular tree) show that the series ε_n and η_n are summable, hence

$$E(d_n) = l\,n + O(1), \qquad \operatorname{var}(d_n) = (1 - l^2)\,n + O(1),$$

which proves that $\lambda^2 = 1 - l^2$.

3.3 Tail of n_R

Let $a = (8/9)^{1/2} < 1$ and $\tau(x)$ be the first hitting time of x if the walk hits x and $\tau(x) = +\infty$ otherwise. Write τ for $\tau(x_0)$ and $1 + Z(x)$ for the degree of x. Call N the subset of B such that x_0 is a vertex never occupied before time zero by the random walk and denote by n_N the first return time to N:

$$N = \{\forall n \le -1,\ x_n \ne x_0\}, \quad n_N(t, \hat{x}) = \inf\{n \ge 1\,;\, S^n(t, \hat{x}) \in N\}.$$

The GP measure of N is strictly positive hence n_N is GP almost surely finite. The next lemma is crucial to our proof.

Lemma 1 – *Let $Z \ge 2$ almost surely. The four following functionals $h(n)$ defined for any integer $n \ge 0$ verify $h(n) = O(a^n)$ as n goes to infinity:*
    (i) $h(n) = GP(\tau = n)$;
    (ii) $h(n) = GP(x_0 = x_n)$;
    (iii) $h(n) = GP(n_N \ge n)$;
    (iv) $h(n) = GP(n_R \ge n^{2+\varepsilon})$ for any $\varepsilon > 0$.
Hence all the powers of n_R and of n_R conditioned by R are integrable.

Proof. We prove (i) by applying Doob's stopping time theorem to a well suited supermartingale. For any $b > 0$ and $n \ge 1$:

$$P(b^{d_{n+1}} \mid d_1, \ldots, d_n, d_n \ne 0, Z(x_n)) = \frac{bZ(x_n) + b^{-1}}{Z(x_n) + 1}\, b^{d_n}.$$

For $b \le 1$ and $Z(x_n) \ge 2$, the fraction written in the right hand member is at most $(2b + b^{-1})/3$. From this point, we set $b = 2^{-1/2}$. One gets the following upper bound:

$$P(b^{d_n+1} \mid d_n \ne 0) \le a\, b^{d_n},$$

so that $(a^{-n} b^{d_n})_n$ is a nonnegative supermartingale under P for $1 \le n \le \tau$. Doob's stopping time theorem yields

$$P(a^{-\tau} \mathbf{1}_{\{\tau < +\infty\}}) \le a^{-1} b = 3/4,$$

so that (i) is true.

By the Markov property of $(x_n)_n$ for a fixed tree t, the series associated to $P(\tau = n)$ and $P(x_0 = x_n)$ are related by

$$\sum_{n \ge 0} P(x_0 = x_n)\, z^n = \left(1 - \sum_{n \ge 1} P(\tau = n)\, z^n\right)^{-1}.$$

By (i), the sum in the right hand member is at most $3/4 < 1$ for $z = a^{-1}$ hence the left hand member is finite for the same value of z, and this proves (ii).

We bound the tail of the law of n_N thanks to the following trivial inclusions:

$$\{n_N \geq n+1\} \subset \{x_n \in \{x_k \,;\, k \leq 0\}\} = \bigcup_{k=0}^{+\infty} \{x_n = x_{-k}\},$$

Integrating term by term, using (ii) and the invariance of G under S, we get (iii).

We now evaluate the tail of the law of n_R. Let $(n_N^k)_k$ be the sequence of successive passage times in N defined by $n_N^1 = n_N$ and

$$n_N^{k+1} = n_N \circ S^{n_N^k}.$$

Each n_N^k is GP almost surely finite and n_R is one of these so that n_R can be large only because one of the return times in N is large or because the rank of n_R in the sequence of the n_N^k's is large. As for the latter, for any $k \geq 1$,

$$P(n_R \geq n_N^{k+1} | n_R \geq n_N^k, x(n_N^k)) = [1 + C(x(n_N^k))]^{-1},$$

where $C(x)$ is the effective conductance of the part of the tree made of the descendants of $x \neq x_0$. This part of t contains a copy of a full binary tree, hence Rayleigh's monotony principle proves that $C(x) \geq 1$ for any $x \neq x_0$ when $Z \geq 2$ almost surely. Hence,

$$P(n_R > n_N^k) \leq 2^{-k}.$$

Finally, n_N^k is the sum of k increments so that it cannot be large without one of these being large. The tail of n_R is bounded as follows:

$$GP(n_R \geq n) \leq GP(n_R > n_N^k) + GP(n_N^k \geq n)$$
$$\leq 2^{-k} + k\,GP(n_N^2 - n_N^1 \geq n/k).$$

The last probability written above is $GP(n_N \geq n/k \,|\, N)$, hence it is bounded by a multiple of $a^{n/k}$. We choose $k = n^{1/2}$ and get the following which is sufficient to prove (iv):

$$GP(n_R \geq n) \leq c\,n^{1/2}\,a^{n^{1/2}}.$$

This concludes the proof of lemma 1.

3.4 Functional central limit theorem

Following an idea due to M. Gordin, see [Gor69], D. Scott gives in [Sco73] a sufficient condition for the sequence of the iterates of a random variable under the action of an ergodic transformation to satisfy a functional central limit theorem. A brief description of this result can be found in Durrett [Dur91, section 7.7]. We recall Scott's result and apply it in our setting.

Let S be an ergodic transformation of a probability space (Ω, Σ, Q), ξ be a square integrable random variable and Σ_0 be a sub-σ-field of Σ such that $\Sigma_0 \subset S^{-1}(\Sigma_0)$. For any (positive or negative) integer n, note $\xi_n = \xi \circ S^n$ and $\Sigma_n = S^{-n}(\Sigma_0)$. Hence the sequence of σ-fields $(\Sigma_n)_n$ is increasing. Scott's key assumption is that the following series converges:

$$\sum_{n \geq 1} \|Q(\xi|\Sigma_{-n})\|_{L^2(Q)} + \|\xi - Q(\xi|\Sigma_n)\|_{L^2(Q)}.$$

Then the $L^2(Q)$ norm of the random variable

$$n^{-1/2} \sum_{k=1}^{n} \xi_k,$$

converges as n goes to infinity to a real $\sigma \geq 0$ and the usual sequence of linear interpolations of the Birkhoff sums of ξ under the action of S satisfies a functional central limit theorem as soon as $\sigma > 0$.

Going back to our setting, we choose $\Omega = B$, $Q = GP$ and Σ_0 the σ-field of the past, that is the σ-field generated by the x_n's for $n \leq 0$. Then the σ-field $S^{-1}(\Sigma_0)$ is generated by the x_n's for $n \leq 1$ so that it contains Σ_0. The first step of the proof is to show that one can replace r_n by the Birkhoff sum associated to a well suited random variable ξ. The second step is to check Scott's condition. The convergence to a normal law which was proved in sections 3.1–3.3 shows then that the variance of $r_n/\sqrt{n}$ converges to a strictly positive limit. The parameter σ of Scott's theorem is strictly positive and this ends the proof of theorem A. We now prove that our problem can be made to satisfy Scott's conditions.

Set $\xi = \mathbf{1}_N - GP(N) = \mathbf{1}(x_0 \notin \{x_n \,;\, n \leq -1\}) - m$, and denote by r'_n the associated Birkhoff sums

$$r'_n = \sum_{k=1}^{n} \xi \circ S^k.$$

One easily sees that $r'_n + nm$ is the cardinal of the set

$$\{x_k \,;\, 1 \leq k \leq n\} \backslash \{x_k \,;\, k \leq 0\},$$

hence $r'_n \leq r_n - n \cdot m$ et $(r_n - n \cdot m) - r'_n$ is uniformly bounded by the cardinal of the set

$$\{x_k \,;\, k \geq 1\} \cap \{x_k \,;\, k \leq 0\}.$$

This set is GP almost surely finite, for example by our bound (ii) of lemma 1 (but in fact by the transience of the random walk). Hence we can replace $r_n - n \cdot m$ by r'_n. Lemma 2 reduces Scott's condition to a control of the tail of n_R.

Lemma 2 – *For any $n > 0$, ξ is Σ_n measurable and*

$$GP(GP(\xi \,|\, \Sigma_{-n})^2) \leq 4\, GP(n_R \geq n).$$

Proof. The sequence $(\Sigma_n)_n$ is increasing and ξ is Σ_0 measurable, hence the first conclusion holds. For the upper bound, we decompose $GP(\xi_n \mid \Sigma_0)$ by comparing n and n_R: we then use the control of the tail of n_R of lemma 1 and the independence of successive slabs to conclude. Decompose ξ_n as

$$\xi_n = \xi_{\sup(n,n_R)} + (\xi_n - \xi_{n_R})\mathbf{1}(n_R \geq n).$$

The difference $|\xi_n - \xi_{n_R}|$ is at most 1. The random variable $\xi_{\sup(n,n_R)}$ is independent of Σ_0 and its mean, using the same relation the other way round, is at most:

$$GP(\xi_{\sup(n,n_R)}) \leq GP(\xi_n) + GP(\mathbf{1}(n_R \geq n)) = GP(n_R \geq n).$$

Hence, we claim that

$$GP(\xi_n \mid \Sigma_0) \leq GP(n_R \geq n) + GP(n_R \geq n \mid \Sigma_0),$$

and this proves lemma 2.

Finally, Scott's condition holds if the following series converges:

$$\sum_n GP(n_R \geq n)^{1/2},$$

and it does in our setting through (iv) of lemma 1. This ends the proof of a functional central limit theorem for the range.

For the distance, one chooses $\xi = [x_0 - x_{-1}]_{x_{-\infty}} - l$ where $[x - y]_s$ is the horodistance between two vertices x and y with respect to the point s of the boundary of the tree (see [LPP95]). We replace $d_n - nl$ by the Birkhoff sum of ξ, causing an error which is uniformly bounded because of the G almost sure transience of the random walk, and Scott's condition is once again satisfied when the sum $\sum_n GP(n_R \geq n)^{1/2}$ is finite.

References

[DS84] P. Doyle and J. Snell. *Random walks and electrical networks.* Math. Association of America, Washington, 1984.

[Dur91] R. Durrett. *Probability: theory and examples.* Wadsworth, 1991.

[Gor69] M. Gordin. The central limit theorem for stationary processes. *Soviet Math. Doklady,* **10**, 1174–1176, 1969.

[JP71] N. Jain and W. Pruitt. The range of transient random walk. *J. Analyse Math.,* **24**, 369–393, 1971.

[LPP95] R. Lyons, R. Pemantle, and Y. Peres. Ergodic theory on Galton-Watson trees: speed of random walk and dimension of harmonic measure. *Ergodic theory Dynam. Systems,* **15**, 593–619, 1995.

[Pi94]   D. Piau. *Quelques inégalités isopérimétriques du mouvement brownien et des marches au hasard sur les graphes.* PhD thesis, Université Claude Bernard (Lyon-I), 1994.

[Pi95]   D. Piau. Deux théorèmes fonctionnels de la limite centrale pour la marche au hasard simple sur un arbre de Galton-Watson supercritique. *C. R. Acad. Sci. Paris,* **321**, série I, 1257–1261, 1995.

[Sco73] D. Scott. Central limit theorems for martingales and for processes with stationary increments using a Skorokhod representation approach. *Adv. Appl. Prob.,* **5**, 119–137, 1973.

Didier Piau
Laboratoire de Probabilités
Université Claude Bernard (Lyon-I)
43, boulevard du 11 novembre 1918
69622 Villeurbanne Cedex (FRANCE)

piau@jonas.univ-lyon1.fr

3. Ultrametric and Algebraic Aspects of Trees

Progress in Probability, Vol. 40

Groupes d'automorphismes et frontières d'arbres: le cas homogène

Francis M. Choucroun

Mathematics Subject Classification: 20B27, 22E35, 22E50, 43A85, 43A90, 60J50

Keywords: Trees, analysis on homogeneous spaces, boundary theory

Abstract. We describe the structure of the automorphism group $\mathcal{G}$ of an homogeneous tree in terms similar to $p-$adic reductive groups. This enables us to make harmonical analysis on it.

Résumé: On décrit la structure du groupe $\mathcal{G}$ des automorphismes d'un arbre homogène $\mathcal{A}$ en des termes analogues à ceux des groupes réductifs $p-$adiques, ceci permet d'en faire l'analyse harmonique, que l'on applique à l'arbre.

1 Motivations

Le groupe $PGL_2(\mathbf{Q}_p)$, des homographies de la droite projective $p-$adique, opère par automorphismes sur un arbre homogène $\mathcal{A}_p$ de type p, ce qui signifie que tous les sommets sont de valence $p+1$; cet arbre, introduit par Bruhat-Tits, popularisé par J.P. Serre dans [S], est constitué des classes d'homothéties de réseaux dans $\mathbf{Q}_p^2$. (Un réseau est un $\mathbf{Z}_p-$module de rang 2.)

Cet arbre est l'analogue du demi-plan de Poincaré $\mathcal{P} = \{z \in \mathbf{C} \mid \operatorname{Im} z > 0\}$ sur lequel le groupe $SL_2(\mathbf{R})$ opère par homographies.

De plus la structure d'espace hyperbolique de $\mathcal{P}$ au sens de Gromov, se retrouvera dans les arbres, que l'on peut considérer comme espaces symétriques discrets de rang un.

Cet exposé est en quelque sorte un guide pour une première lecture du mémoire S.M.F. [Ch-1], où l'on trouvera dans un cadre plus général les démonstrations, et auquel les références non précisées se rapportent.

2 Géométrie de l'arbre

Dans ce numéro on considérera un arbre $\mathcal{A}$ (combinatoire) composé d'un ensemble de sommets S, auquel on pourra identifier l'arbre, et d'arêtes A (non orientées), qui est localement fini, c'est à dire dont tous les sommets ont un ordre fini, l'ordre d'un sommet étant le nombre d'arêtes auxquelles il appartient. Dans un arbre il y a une distance naturelle définie sur les sommets, pour laquelle les arêtes sont constituées de deux sommets à distance 1, il y a des géodésiques, et le segment $[a,b] = \{c \in \mathcal{A} \mid d(c,a) + d(c,b) = d(a,b)\}$ est une géodésique de longueur $d(a,b)$.

Parmi les demi-géodésiques, appelées encore rayons géodésiques, qui sont les images isométriques de $\mathbf{N}$, on a une relation d'équivalence:

$$(x_n)_{n \geq 0} \sim (y_n)_{n \geq 0} \iff \exists p \mid x_n = y_{n+p} \text{ pour } n \text{ assez grand.}$$

On appelle bout un élément de l'ensemble quotient, que l'on note Ω; on écrira $[a,\omega]$ la demi-géodésique appartenant au bout ω, telle que $x_0 = a$, que l'on dira issue du sommet a.

On considère aussi des géodésiques qui sont les images isométriques de $\mathbf{Z}$, et étant donnés deux bouts distincts ω et ω', on notera $[\omega,\omega']$ la géodésique $(x_n)_{n\in\mathbf{Z}}$ telle que $(x_n)_{n\geq 0} \in \omega'$ et $(x_{-n})_{n\geq 0} \in \omega$.

Etant donnés trois bouts distincts ω,ω',ω'', les trois géodésiques qu'ils forment $[\omega,\omega']$, $[\omega,\omega'']$ et $[\omega',\omega'']$ se rencontrent en un unique sommet, appelé carrefour des trois bouts et noté $(\omega,\omega',\omega'')$; c'est aussi la projection d'un bout sur la géodésique déterminée par les deux autres.

L'espace discret $\mathcal{A}$ se compactifie en $\overline{\mathcal{A}} = \mathcal{A} \cup \Omega$, dont la topologie admet une base d'ouverts constituée, d'une part des ensembles $V(a,b) = \{x \in \overline{\mathcal{A}} \mid b \in [a,x]\}$ attachés aux couples $a,b \in \mathcal{A}, a \neq b$, d'autre part des ensembles réduits à un sommet. Pour cette topologie $\mathcal{A}$ reste discret et partout dense dans $\overline{\mathcal{A}}$ qui est compact.

On appelle automorphisme de l'arbre $\mathcal{A}$ une bijection compatible avec la structure, c'est à dire respectant l'ensemble des arêtes. Dans le cas d'un arbre combinatoire considéré ici, il y a identité entre automorphisme et isométrie de l'arbre.

Le groupe $G = \mathrm{Aut}(\mathcal{A})$ des automorphismes de l'arbre, muni de la topologie de la convergence simple dans l'espace discret $\mathcal{A}$, est un groupe localement compact totalement discontinu, pour lequel les fixateurs des sommets sont ouverts compacts.

Tout automorphisme se prolonge par continuité en bijection de $\overline{\mathcal{A}}$ et de Ω.

Un invariant permet de décrire les bijections de Ω que l'on obtient ainsi.

On note $(a,b;\omega,\omega')$, pour deux sommets a,b d'une géodésique orientée $[\omega,\omega']$, paramétrée par $(x_n)_{n\in\mathbf{Z}}$, l'entier relatif k, tel que si $a = x_p$ on ait $b = x_{p+k}$. On définit alors pour quatre bouts $(\omega_i)_{i=1,\ldots,4}$ tels que les carrefours $a = (\omega_1,\omega_3,\omega_4)$ et $b = (\omega_2,\omega_3,\omega_4)$ soient définis, l'invariant $\iota(\omega_1,\omega_2;\omega_3,\omega_4) := (a,b;\omega_3,\omega_4)$.

On vérifie sans peine que ι est invariant par tout automorphisme de l'arbre, et que dans un arbre sans sommet d'ordre 1 ou 2 toute bijection de l'ensemble des bouts qui respecte ι provient d'un automorphisme de l'arbre.

Dans le cas de l'arbre $\mathcal{A}$ attaché à $PGL_2(\mathbf{Q}_p)$, le groupe des automorphismes G contient $PGL_2(\mathbf{Q}_p)$. L'espace des bouts de $\mathcal{A}$ s'identifie à la droite projective $\mathbf{P}^1(\mathbf{Q}_p)$, et $PGL_2(\mathbf{Q}_p)$ y opère par homographies, et on vérifie, notamment en utilisant la description donnée dans [Ch-2], que l'on a $\iota(\omega_1,\omega_2;\omega_3,\omega_4) = v_p[\omega_1,\omega_2;\omega_3,\omega_4]$, où $[?,?;?,?]$ désigne le birapport, et v_p la valuation $p-$adique. Ceci permet de comparer ces deux groupes. Considérés comme groupes de bijections de la droite projective $\mathbf{P}^1(\mathbf{Q}_p)$, ils sont caractérisés par les invariants qu'ils respectent. Pour $PGL_2(\mathbf{Q}_p)$ cet invariant est le birapport de quatre bouts, tandis que pour G c'est la valuation $p-$adique de ce même birapport.

On peut étudier divers types d'arbres. On appelle arbre numéroté (ou coloré), un arbre muni d'une application surjective dans un ensemble I de numéros, appelée numérotation, et automorphisme d'un arbre numéroté, un automorphisme de l'arbre qui commute avec la numérotation. (L'ensemble I peut être réduit à un élément!)

On appellera arbre régulier, un arbre numéroté dans lequel, étant donné deux numéros $i, j \in I$ il existe une constante $m(i, j)$, telle que pour tout sommet x de numéro i le nombre d'arêtes de la forme $[x, y]$, où y est de numéro j est égal à $m(i, j)$. Parmi les arbres réguliers, on peut considérer les arbres semi-homogènes, où $I = \{1, 2\}$ et $m(1, 1) = m(2, 2) = 0$. On dira qu'un arbre semi-homogène $\mathcal{A}_{q_1, q_2}$ est de type (q_1, q_2) si $m(1, 2) = q_1 + 1$, et $m(1, 2) = q_2 + 1$.

A un arbre homogène $\mathcal{A}$ de type q, dont S est l'ensemble des sommets et A l'ensemble des arêtes, on peut associer un arbre semi-homogène $\mathcal{A}_{q,1}$ de type $(q, 1)$, en rajoutant comme sommets de numéro 2 les "milieux" des arêtes, c'est à dire dont l'ensemble des sommets de numéro 1 (resp. 2) est S (resp. A) et en considérant comme arête du nouvel arbre les couples consitués d'une arête de $\mathcal{A}$ et d'un sommet de cette arête de $\mathcal{A}$ (arête orientée). On vérifie que les bouts de $\mathcal{A}$ et $\mathcal{A}_{q,1}$ s'identifient, de même que les groupes d'automorphismes d'arbre de $\mathcal{A}$ et d'arbre numéroté $\mathcal{A}_{q,1}$. Les fixateurs d'un sommet a de numéro 2 a une structure intéressante: c'est le stabilisateur de l'arête a de $\mathcal{A}$, qui contient le fixateur de a comme sous-groupe d'indice 2.

On peut aussi transformer un arbre homogène $\mathcal{A}$ de type q en arbre semi-homogène $\mathcal{A}_{q,q}$ de type (q, q) en le numérotant par la parité de distance à un sommet donné. L'espace des bouts reste inchangé, mais le groupe des automorphismes de l'arbre $\mathcal{A}_{q,q}$ est un sous-groupe d'indice 2 de celui des automorphismes de $\mathcal{A}$.

Enfin, on vérifie que le groupe G des automorphismes d'un arbre homogène $\mathcal{A}$ est transitif sur l'espace de ses bouts.

On peut montrer [1.6.1-1.6.2] que cette condition caractérise essentiellement parmi les arbres réguliers sans sommet d'ordre 1 ou 2, les arbres homogènes ou semi-homogènes, que l'on appelle arbres de Bruhat-Tits, dont la structure des groupes d'automorphismes est voisine et qui sont étudiés dans [Ch-1]. Voir aussi [Ca-1] et [Ca-2].

3 Structure du groupe d'automorphismes d'un arbre homogène [1.5]

Désormais, on considérera un arbre homogène $\mathcal{A}$ d'ordre $q > 1$ fixé, dont on note G le groupe d'automorphismes.

Cette section décrit la structure de G en termes analogues à celle de $PGL_2(\mathbf{Q}_p)$.

Le point clef est que le groupe G est alors doublement transitif au sens qu'étant donnés deux segments $[x, x']$ et $[y, y']$ de même longueur, il existe un élément $g \in G$ tel que $g(x) = y$ et $g(x') = y'$: cette condition est équivalente à la transitivité de G sur chaque sphère.

Le groupe G possède un caractère d'ordre 2, défini par $\varepsilon(g) = (-1)^{d(x,g(x))}$, (cette définition ne dépend pas du sommet x de l'arbre). On note $G^+ = \ker \varepsilon$, c'est un sous-groupe d'indice 2 de G.

Soit $\omega \in \Omega$, on notera B_ω le sous-groupe des éléments de G qui fixe le bout ω.

L'image par un élément $b \in B_\omega$ d'une demi-géodésique $(x_n) \in \omega$ restant dans ω, il existe un entier $\nu_\omega(b)$ qui ne dépend que de b, tel que $b(x_n) = x_{n+k}$ avec $k = \nu_\omega(b)$ pour n assez grand. De plus ν_ω est un homomorphisme de B_ω sur $\mathbf{Z}$, dont on note $N_\omega = \ker \nu_\omega$.

Pour continuer l'étude de G, on choisit une géodésique pointée et orientée que l'on écrit (x_n), et dont les bouts seront notés ∞ et $-\infty$ avec $(x_n)_{n>0} \in \infty$ et $(x_{-n})_{n>0} \in -\infty$.

On notera $B = B_\infty$, $B^- = B_{-\infty}$, $N = N_\infty$, et $K = G_{x_0}$ qui est un sous-groupe compact maximal, on appelera sous-groupe d'Iwahori le fixateur d'une arête, et on posera $I = G_{x_0} \cap G_{x_1}$. Le stabilisateur de l'arête $\{x_0, x_1\}$ est le normalisateur $N(I)$ de I, dans lequel I est un sous-groupe d'indice 2.

On remarquera que l'espace homogène G/K s'identifie à l'ensemble des sommets de l'arbre, tandis que $G/N(I)$ s'identifie à celui des arêtes, tandis que G/I s'identifie à l'ensemble $\hat{A}$ des arêtes orientées (avec une source et un but).

On désigne enfin par $N_k = \{n \in G \mid n|_{[x_k,\infty]} = Id\} \subset N$, c'est un groupe compact, et $H = \bigcap_{k \in \mathbf{Z}} N_k = \{n \in N \mid \forall k \ n(x_k) = x_k\}$, qui est le fixateur de la géodésique $[-\infty, \infty]$.

On définit une sorte de valuation sur $N - H$ par

$$v(n) = \inf\{k \mid n \in N_k\} = \inf\{k \mid n(x_k) = x_k\}.$$

Proposition 1. *1) Il existe dans G des éléments w, τ tels que $\forall n$, $w(x_n) = x_{-n}$, $\tau(x_n) = x_{n+1}$.*
2) On a la décomposition de Cartan

$$G = \bigsqcup_{n \in \mathbf{N}} K\tau^n K, \text{ et } g \in K\tau^n K \iff d(g(x_0), x_0) = n.$$

Les classes KgK sont invariantes par $g \mapsto g^{-1}$, ainsi (G, K) est une paire de Gelfand.
3) On a

$$B = \tau^{\mathbf{Z}} N \text{ et } K \cap N = N_0$$

4) On a la décomposition d'Iwasawa

$$G = KB = K\tau^{\mathbf{Z}} N$$

5) On a la décomposition $K = I \sqcup IwI$, et la décomposition d'Iwasawa peut être précisée en $G = IB \sqcup IwB$.

6) On a $I = (I \cap B).(I \cap B')$, avec $(I \cap B) \cap (I \cap B') = H$.
7) On a la décomposition de Bruhat

$$G = NwB \sqcup B$$

et $n_1 wB \cap n_2 wB \neq \emptyset \iff n_1 H = n_2 H$.
8) L'indice $[N_k : N_{k-1}]$ vaut q.
9) (Relation entre les décompositions de Cartan et d'Iwasawa) Pour $n \in N$, avec $v(n) = k$, on a

$$\tau^l n \in K \tau^{m(k,l)} K$$

avec $m(k,l) = |l|$ si $k \leq \max(0,-2l)$ et $m(k,l) = l + k$ si $k \geq \max(0,-2l)$.
10) Soit $\widetilde{W} = \tau^{\mathbf{Z}} \cup \tau^{\mathbf{Z}} w$. Il stabilise $[-\infty, \infty]$, et y opère comme le groupe diédral infini D_∞, et on a

$$G = \bigsqcup_{g \in \widetilde{W}} IgI.$$

Remarques. Le point 2) décrit la position de deux sommets de l'arbre par l'élément de $\mathbf{N}$ qui est leur distance, de même le point 10 décrit celle des arêtes orientées par un élément du groupe diédral D_∞ réalisé ensemblistement par $\widetilde{W}$.

Comme (G, K) est une paire de Gelfand, $\mathcal{C}_c(K \backslash G / K)$, l'algèbre pour la convolution des fonctions bi-invariantes par K à support compact sur G, sera commutative.

La transitivité de G à la fois sur l'ensemble des sommets et sur l'ensemble des bouts permet d'interpréter la décomposition $G = BK = KB$ comme la transitivité de K sur Ω, et plus généralement celle de G_x, de même que celle de B, voire celle de B_ω, sur $\mathcal{A}$.

Exemple. On considère l'arbre $\mathcal{A}_p$ attaché à $PGL_2(\mathbf{Q}_p)$, dans lequel on définit x_n comme la classe du réseau $\mathbf{Z}_p \times p^n \mathbf{Z}_p$, on choisit alors la géodésique $(x_n) = [-\infty, \infty]$, et dans l'identification de Ω avec $\mathbf{P}^1(\mathbf{Q}_p)$, le bout ∞ est la droite $\mathbf{Q}_p \times \{0\}$ et le bout $-\infty$ est $\{0\} \times \mathbf{Q}_p$.

Pour un sous-groupe $X \subset G$, on notera $\widetilde{X} = X \cap PGL_2(\mathbf{Q}_p)$ et $\overline{X}$ le sous-groupe de $GL_2(\mathbf{Q}_p)$ qui se projette sur $\widetilde{X}$.

Ainsi $\overline{B}$ est le sous groupe triangulaire supérieur de $GL_2(\mathbf{Q}_p)$, et on a

$$\overline{N} = \left\{ \begin{pmatrix} a & b \\ 0 & d \end{pmatrix} \mid |\tfrac{a}{d}| = 1 \right\} \supset \overline{N_n} = \left\{ \begin{pmatrix} a & b \\ 0 & d \end{pmatrix} \mid |\tfrac{a}{d}| = 1, |\tfrac{b}{d}| \leq p^{-n} \right\},$$

de sorte que $\overline{H} = \left\{ \begin{pmatrix} a & 0 \\ 0 & d \end{pmatrix} \mid |\tfrac{a}{d}| = 1 \right\}$, et pour $n = \begin{pmatrix} a & b \\ 0 & 1 \end{pmatrix} \in N$ on a $p^{v(n)} = |b|$.

On vérifie que $\widetilde{K} = PGL_2(\mathbf{Z}_p)$.

Enfin on pourra choisir $\tau = \begin{pmatrix} 1 & 0 \\ 0 & p \end{pmatrix}$, et $w = \begin{pmatrix} 0 & 1 \\ 1 & 0 \end{pmatrix}$.

L'analogie avec les groupes réductifs qui s'exprime par la formule classique

$$\begin{pmatrix} 1 & 0 \\ -t^{-1} & 1 \end{pmatrix} \begin{pmatrix} 1 & t \\ 0 & 1 \end{pmatrix} \begin{pmatrix} 1 & 0 \\ -t^{-1} & 1 \end{pmatrix} = \begin{pmatrix} 0 & t \\ -t^{-1} & 0 \end{pmatrix}$$

se retrouve ci-dessous.

Proposition 2. *Pour tout $n \in N - H$, il existe $n', n'' \in N - H$ tels que*

$$wn'wnwn'' = \tau^k$$

avec $k = v(n) = -v(n') = -v(n'')$.

Etant donnés deux sommets $x, y \in \mathcal{A}$ et un bout $\omega \in \Omega$, on définit la position de x, y par rapport à ω par

$$< x, y; \omega >= d(y, z) - d(x, z) \text{ pour } z \in [x, \omega] \cap [y, \omega].$$

On vérifie que la différence $d(y, z) - d(x, z)$ ne dépend pas du sommet choisi $z \in [x, \omega] \cap [y, \omega]$, qu'elle définit un cocycle, appelé par certains auteurs cocycle de Buseman par analogie avec les espaces symétriques, et que la relation $\mathcal{R}_\omega$ sur $\mathcal{A}$ définie par

$$x\mathcal{R}_\omega y \Longleftrightarrow < x, y; \omega >= 0,$$

est une relation d'équivalence, dont les classes sont appelées horocycles issus de ω.

On montre enfin que la relation $< x, y; \omega >= 0$ équivaut à $\exists n \in N_\omega$ tel que $y = n(x)$. Ainsi les horocycles issus de ω sont les orbites de N_ω dans l'arbre.

4   Mesures [1.7-1.8-1.9]

Le groupe G admet une mesure de Haar dg, normalisée par sa restriction au sous-groupe ouvert compact K que l'on choisi de volume 1. C'est un groupe unimodulaire, ainsi les groupes G_x, qui sont des conjugués de K sont de volume 1.

Le groupe B n'est pas unimodulaire, on vérifie que son module δ est donné par la formule $\delta(b) = q^{\nu(b)}$, d'où l'on déduit

$$\int_G f(g)d(g) = \int_K \sum_{l \in \mathbf{Z}} \int_N f(k\tau^l n)q^l dk dn.$$

Les groupes G_x opèrent continûment et transitivement sur Ω : ceci permet pour tout sommet x de considérer Ω comme espace homogène sous G_x, isomorphe à $G_x/(G_x \cap B)$ qui est muni d'une mesure de Haar invariante, d'où une mesure de probabilité ν_x sur Ω.

En notant $\Omega_{x,y} = \Omega \cap V_{x,y}$, et $S(x, n) = \{z \mid d(x, z) = n\}$, on a pour $d(x, y) = n > 0$

$$\nu_x(\Omega_{x,y}) = \frac{1}{\mathrm{Card}(S(x, n))} = \frac{1}{(q + 1)q^{n-1}}.$$

On appelle noyau de Poisson K la dérivée de Radon-Nikodym

$$K(x, y; \omega) = \frac{d\nu_y(\omega)}{d\nu_x(\omega)}.$$

On vérifie que $K(x, y; \omega) = q^{<x,y;\omega>}$, de sorte que, comme pour le demi-plan de Poincaré, le noyau de Poisson est constant sur les horocycles.

On a la formule $\frac{1}{|S(x,n)|} \sum_{y \in S(x,n)} K(x, y; \omega) = 1$, ce qui compte tenu de la définition des noyaux de Poisson comme quotients, en montre l'harmonicité.

Le fixateur K^n de la boule $B(x_0, n) = \{y \mid d(x_0, y) \leq n\}$ est un sous-groupe ouvert compact de K, et l'ensemble de ces groupes constitue une base de voisinage de l'unité dans K et dans G.

Par ailleurs, pour chaque n on définit sur Ω la tribu $\mathcal{B}^n$ engendrée par les ensembles $\Omega_{x_0,y}$ pour $y \in S(x_0, n)$.

Ayant noté T l'action sur les fonctions f définies sur Ω, déduite de celle de G, c'est à dire $T(g)(f)(\omega) = f(g(\omega))$, ε_n la fonction caractéristique de K^n normalisée par sa mesure, et $*$ la convolution à gauche dans le K-espace Ω, on a

Proposition 3. *Soit $f \in L^2(\Omega)$, f est $\mathcal{B}^n$-mesurable $\Longleftrightarrow$ pour tout $k \in K^n$ on a $T(k)f = f$.*

On en déduit que l'espérance conditionnelle $E^{\mathcal{B}^n}$ est donnée par

$$E^{\mathcal{B}^n}(f) = \varepsilon_n * f.$$

5   Représentations [2.1-2.2-2.5]

L'objet de ce numéro est de construire et d'étudier des représentations de G qui interviendront dans des décompositions spectrales des espaces

$$L^2(G/K), \quad L^2(G/N(I)), \quad \text{et} \quad L^2(G/I).$$

Ces représentations seront définies de façon analogue à la série principale de $SL_2(\mathbf{R})$ que l'on peut trouver dans le chapitre III de [L].

Etant donné $\lambda \in \mathbf{C}^*$, la formule $\chi_\lambda(b) = (\sqrt{q}\lambda)^{\nu(b)}$ définit un caractère χ_λ de B.

On définit alors

$$V^\lambda = \{f \in \mathcal{C}^\infty(G) | \text{pour } g \in G, \ b \in B \ \ f(gb) = f(g)\chi_\lambda^{-1}(b)\},$$

où on note $\mathcal{C}^\infty(X)$ l'espace des fonctions complexes localement constantes sur un espace X totalement discontinu, et $\mathcal{C}_c^\infty(X)$ son sous-espace constitué des fonctions à support compact.

Le groupe G opère sur V^λ par la représentation régulière gauche, que l'on note π^λ, c'est à dire que l'on a, pour $f \in V^\lambda$ et $g \in G$, $\pi^\lambda(f)(x) = f(g^{-1}x)$.

Par ailleurs, on définit sur V^λ une forme hermitienne par

$$< f, g >= \int_K f(k)\bar{g}(k)\, dk,$$

dont on notera $\mathcal{H}^\lambda$ le complété.

On montre que les opérateurs $\pi^\lambda(g)$ sont bornés pour cette forme, et se prolongent ainsi en opérateurs sur l'espace de Hilbert $\mathcal{H}^\lambda$, que l'on note $\Pi^\lambda(g)$.

On peut aussi réaliser $\mathcal{H}^\lambda$ comme l'espace des classes de fonctions sur G qui vérifient $f(gb) = f(g)\chi_\lambda^{-1}(b)$ pour $g \in G$ et $b \in B$, et telles que $f|_K \in L^2(K)$: ainsi la représentation Π^λ peut s'identifier à la représentation régulière de G dans $\mathcal{H}^\lambda$.

On remarquera que les espaces $\mathcal{H}^\lambda$ et V^λ possèdent comme sous-espace invariant par K la droite $\mathbf{C}1_\lambda$, où la fonction 1_λ est définie par $1_\lambda(kb) = \chi_\lambda^{-1}(b)$.

On remarquera que la restriction à K de Π^λ (resp. π^λ) est la représentation régulière de K dans $L^2(K/(K \cap B))$ (resp. $\mathcal{C}^\infty(K/(K \cap B))$.)

On va construire des représentations dans les espaces de fonctions sur Ω qui seront liées aux précédentes.

On définit, pour $s \in \mathbf{C}$, la représentation ω^s de G dans $\mathcal{C}^\infty(\Omega)$ par

$$(\omega^s(g)f)(\omega) = K^s(g(x_0), x_0; \omega)f(g^{-1}(\omega)),$$

dans cette formule $K^s(.,.;.)$ désigne, bien entendu, la puissance s−ième de $K(.,.;.)$, on vérifie que c'est une représentation et que s n'intervient que par q^s.

La représentation ω^s est bornée dans $L^2(\Omega)$ et se prolonge en une représentation de G que l'on note Ω^s. Enfin, on vérifie que la restriction à K de ω^s (resp. Ω^s) est la représentation régulière de K dans $\mathcal{C}^\infty(\Omega)$ (resp. $L^2(\Omega)$); on démontre plus précisement.

L'application $g \mapsto g(\infty)$ définit une bijection de $G/B = K/(K \cap B)$ sur Ω, qui induit des bijections entre d'une part $\mathcal{C}^\infty(K/(K \cap B))$ et $\mathcal{C}^\infty(\Omega)$, et d'autre part $L^2(K/(K \cap B))$ et $L^2(\Omega)$, compatibles avec la structure de K−espaces, ce qui permet de les identifier. On montre que ces identifications entre K−espaces sont compatibles aux représentations de G, de sorte qu'on obtient les équivalences canoniques

$$\text{Pour}\sqrt{q}\lambda = q^s, \text{ on a } \omega^s \sim \pi^\lambda \text{ et } \Omega^s \sim \Pi^\lambda.$$

Proposition 4. *1) Pour $\lambda \neq \pm\sqrt{q}^{\pm 1}$ les représentations π^λ (resp. Π^λ) sont algébriquement (resp. topologiquement) irréductibles.*

2) La représentation Π^λ admet comme contragrédiente la représentation $\Pi^{\lambda'}$ pour $\lambda' = \bar{\lambda}^{-1}$.

Pour $|\lambda| = 1$ les représentations Π^λ sont unitaires. (Séries principales unitaires.)

3) On a $\pi^{-\lambda} = \pi^\lambda \otimes \varepsilon$, et $\Pi^{-\lambda} = \Pi^\lambda \otimes \varepsilon$, avec ε le caractère d'ordre deux de G.

4) Pour $\lambda \neq \pm\sqrt{q}^{\pm 1}$ les représentations π^λ et $\pi^{\lambda^{-1}}$ sont équivalentes. Si de plus $|\lambda| = 1$ les représentations Π^λ et $\Pi^{\lambda^{-1}}$ sont unitairement équivalentes.

5) On suppose $|\lambda| \neq 1$: les représentations π^λ sont unitarisables si et seulement si $|\lambda| \in]\frac{1}{\sqrt{q}}, \sqrt{q}[$. (Séries complémentaires)

Pour $\lambda = \sqrt{q}$, le caractère χ_λ est le module δ, ainsi $\int_K f(k)\, dk$ définit une forme linéaire invariante par G. On appelle représentation de Steinberg le sous-G-module

$$St = \{f \in \mathcal{H}^\lambda \mid \int_K f = 0\} \quad \text{ou} \quad st = \{f \in V^\lambda \mid \int_K f = 0\}.$$

On peut montrer que st (resp. St) est algébriquement (resp. topologiquement) irréductible (prop. 2.6.1). On montre que les coefficients de st sont de carré intégrables sur G (prop. 2.7.5), ainsi st peut être réalisé dans $L^2(G)$, dont on notera $\mathbf{St}$ l'adhérence. Enfin on vérifie que les $G-$modules quotients π^λ/st, et Π^λ/St sont les modules identité.

Par contre si $\lambda = \sqrt{q}^{-1}$, le caractère χ_λ est trivial, la représentation π^λ (resp. Π^λ) est la représentation régulière dans $\mathcal{C}^\infty(G/B)$ (resp. $L^2(G/B)$), qui possède l'ensemble $\mathbf{C}1$ des fonctions constantes, comme un sous-espace invariant.

On notera les G-modules quotients $St' = \mathcal{H}^\lambda/\mathbf{C}1$ et $st' = V^\lambda/\mathbf{C}1$, que l'on appelera encore représentation de Steinberg, car on montre que st et st' sont algébriquement équivalents.

On va donner une description géométrique de la représentation de Steinberg $\mathbf{St}$ (prop. 3.7.3). Soit $\widehat{A} = G/I$ l'ensemble des arêtes orientées de l'arbre. Pour tout sommet x de l'arbre et pour toute fonction f définie sur $\widehat{A}$, on note $M_x(f)$ la moyenne de f sur l'ensemble des arêtes orientées dont l'origine est x. On notera le changement d'orientation dans $\widehat{A}$, obtenu en échangeant pour chaque arête source et but, par $a \mapsto \overline{a}$.

On considère alors

$$V^0 = \{f \in L^2(\widehat{A}) \mid \forall a \in \widehat{A} \;\; f(\overline{a}) = -f(a), \;\; \text{et } \forall x \in \mathcal{A} \;\; M_x(f) = 0\},$$

c'est un sous-espace de $L^2(G/I) \subset L^2(G)$ invariant par G, et $\mathbf{St}$ peut y être identifié.

6   Fonctions radiales [3.1-3.2]

On va commencer à étudier les algèbres de fonctions radiales sur l'arbre, c'est à dire qui ne dépendent que la distance $d(x_0, x)$. Compte tenu de la décomposition de Cartan, on les considère comme des fonctions sur G biinvariantes par K. On considèrera les algèbres, pour la convolution, $\mathcal{C}_c(K\backslash G/K)$ et $L^1(K\backslash G/K)$. Elles sont commutatives.

Pour cette étude on sera amené à définir, pour fonction $f \in L^1(K\backslash G/K)$, sa transformèe de Satake $\tilde{f}$ définie sur $K\backslash G/N = \tau^{\mathbf{Z}}$ par

$$\tilde{f}(\tau^p) = \delta^{\frac{1}{2}}(\tau^p) \int_N f(\tau^p n)\, dn = q^{\frac{p}{2}} \int_N f(\tau^p n)\, dn.$$

On montre que si $f \in \mathcal{C}_c(K\backslash G/K)$, sa transformée de Satake est à support fini, et vérifie $\tilde{f}(\tau^{-p}) = \tilde{f}(\tau^p)$, et que la transformation de Satake est une bijection de $\mathcal{C}_c(K\backslash G/K)$ sur l'ensemble des fonctions sur $\tau^{\mathbf{Z}}$ qui vérifient ces conditions.

Pour $f \in L^1(K\backslash G/K)$, on considère l'opérateur $\Pi^\lambda(f) = \int_G \Pi^\lambda(g)f(g)\,dg$: il est de rang un, son image est la droite $\mathbf{C}1_\lambda$ des vecteurs invariants par K, sur laquelle il opère par un scalaire, que l'on note $\varphi_\lambda(f)$. On vérifie que $f \mapsto \varphi_\lambda(f)$ est un caractère de l'algèbre $L^1(K\backslash G/K)$, c'est à dire que l'on a $\varphi_\lambda(f \star g) = \varphi_\lambda(f)\varphi_\lambda(g)$, et on montre que tous les caractères cette algèbre sont ainsi obtenus.

En calculant $\varphi_\lambda(f)$, on montre que c'est la transformée de Fourier-Laplace de $\tilde{f}$, donnée par $\varphi_\lambda(f) = \sum_{p\in\mathbf{Z}} \lambda^{-p}\tilde{f}(\tau^p)$. On pourra noter de façon simplifiée $\varphi_\lambda(f) = \hat{f}(\lambda)$, quand on la considérera comme transformée de Fourier sphérique de f.

Ainsi l'analyse harmonique des fonctions radiales sera la conséquence du calcul de la transformation de Satake et de son inversion.

On introduit la fonction c_q d'Harish-Chandra définie sur $\mathbf{C}^*$ par

$$c_q(\lambda) = \frac{1 - \frac{1}{q^2}\lambda^{-2}}{1 - \lambda^{-2}},$$

et C_q la fonction sur le groupe $\tau^{\mathbf{Z}}$, dont c_q est la transformée de Fourier-Laplace. C'est à dire donnée par $C_q(\tau^n) = a_n$, pour $c_q(\lambda) = \sum_{n\in\mathbf{N}} a_n\lambda^{-n}$.

Proposition 5. *La restriction à $\tau^{\mathbf{N}}$ de la transformée de Satake $\tilde{f}$ d'une fonction f de $\mathcal{C}_c(K\backslash G/K)$ est donnée par*

$$\tilde{f}|_{\tau^{\mathbf{N}}} = (\delta^{\frac{1}{2}}f) \star C_q,$$

où $\star$ désigne la convolution sur le groupe $\tau^{\mathbf{Z}}$.

On définit les fonctions sphériques élémentaires ϖ_λ comme coefficients de la représentation Π^λ par $\varpi_\lambda(g) = <\Pi^\lambda(g)\mathbf{1}_\lambda, \mathbf{1}_{\lambda'}>$, où on aura choisi $<f,g> = \int_K f(k)\bar{g}(k)\,dk$ comme accouplement entre $\mathcal{H}^\lambda$ et sa contragrédiente $\mathcal{H}^{\lambda'}$.

On peut expliciter les fonctions sphériques élémentaires ϖ_λ en utilisant la transformation de Satake, on obtient pour $p \geq 0$

$$\varpi_\lambda(\tau^p) = \frac{1}{1 + \frac{1}{q}} q^{-\frac{p}{2}} \left[c_q(\lambda)\lambda^p + c_q(\lambda^{-1})\lambda^{-p}\right].$$

Les caractères de $L^1(K\backslash G/K)$ s'expriment alors comme valeurs à l'origine e d'un produit de convolution sur G au moyen de la formule

$$\varphi_\lambda(f) = (f \star \varpi_\lambda)(e).$$

7 Formules d'inversion de Fourier et de Plancherel sphériques [3.3]

On note γ le cercle unité orienté dans le sens direct, γ_1^+ son intersection avec le demi-plan supérieur, muni de la mesure de Haar $d^*\lambda = \frac{1}{2i\pi}\frac{d\lambda}{\lambda}$.

Proposition 6. *1) Pour une fonction $f \in L^1(K\backslash G/K)$, on a*

$$f(e) = (1 + \frac{1}{q}) \int_{\gamma_1^+} \hat{f}(\lambda)\big|\frac{1}{c_q(\lambda)}\big|^2 d^*\lambda.$$

2) Pour une fonction $f \in L^2(K\backslash G/K)$, et si $p \geq 0$ on a

$$\|f\|_2^2 = (1 + \frac{1}{q}) \int_{\gamma_1^+} |\hat{f}|^2(\lambda)\big|\frac{1}{c_q(\lambda)}\big|^2 d^*\lambda,$$

$$\text{et } f(\tau^p) = (1 + \frac{1}{q}) \int_{\gamma_1^+} \hat{f}(\lambda)\varpi_\lambda(\tau^p)\big|\frac{1}{c_q(\lambda)}\big|^2 d^*\lambda.$$

Le point 1) est la formule d'inversion de Fourier, que l'on établit d'abord pour les fonctions à support compact.

On notera dm la mesure sur γ_1^+ égale à $dm(\lambda) = (1 + \frac{1}{q})\big|\frac{1}{c_q(\lambda)}\big|^2 d^*\lambda$.

Le point 2) se déduit de 1) pour les fonctions à support compact, permettant de définir ensuite $\hat{f}(\lambda)$ comme élément de $L^2(\gamma_1^+, dm)$ pour f dans $L^2(K\backslash G/K)$, et on montre enfin que l'application $f \mapsto \hat{f}$ est un isomorphisme d'espace de Hilbert.

8 Décompositions spectrales [3.5–3.6]

Pour décomposer la représentation régulière de G dans $L^2(G/K)$, on utilise son commutant qui est $L^2(K\backslash G/K)$, auquel on applique la théorie spectrale des fonctions radiales établie ci-dessus. On obtient

Théorème 7. *[3.4.1] La représentation régulière de G dans $L^2(G/K)$, qui s'identifie à l'espace $L^2(\mathcal{A})$ des fonctions de carré sommable sur les sommets de l'arbre, est égale à l'intégrale hilbertienne des $\mathcal{H}^\lambda$ pour la mesure dm de support γ_1^+, et qui y vaut $(1 + \frac{1}{q})\big|\frac{1}{c_q(\lambda)}\big|^2 d^*\lambda$.*

$$L^2(G/K) = \int_{\gamma_1^+}^{\oplus} \mathcal{H}^\lambda dm(\lambda),$$

où G opère dans $\mathcal{H}^\lambda$ par Π^λ. De plus si $f \in \mathcal{C}_c^\infty(G/K)$, son image $(f_\lambda)_{\lambda\in\gamma_1^+}$ est donnée par $f_\lambda = \pi^\lambda(f)(\mathbf{1}_\lambda)$.

On désigne par $T^0 = \varepsilon \otimes \mathbf{St}$, on vérifie qu'elle est de carré intégrable, et qu'elle possède une droite de vecteurs invariants par le sous-groupe compact maximal $N(I)$ de G. En faisant comme dans [Ch-1] la théorie pour les arbres semi-homogène, on obtient la décomposition spectrale de l'espace $L^2(A)$ des fonctions de carré sommable sur les arêtes (non orientées) de l'arbre. Pour cela on notera que $G = N(I)B$, de sorte que les éléments de $\mathcal{H}^\lambda$ sont déterminés par leur valeurs sur $N(I)$, et on définira $\psi_\lambda \in \mathcal{H}^\lambda$ comme la fonction constante sur $N(I)$ qui y vaut $(\mathrm{mes}(N(I)))^{-1} = \frac{2}{q+1}$.

Théorème 8. *La représentation régulière de G dans $L^2(G/(N(I))$, qui s'identifie à l'espace $L^2(A)$, est la somme directe de la représentation T^0, et d'une partie continue qui est intégrale hilbertienne des $\mathcal{H}^\lambda$ pour la mesure dm de support γ_1^+.*

$$L^2(G/K) = \int_{\gamma_1^+}^{\oplus} \mathcal{H}^\lambda dm(\lambda) \oplus T^0,$$

où G opère dans $\mathcal{H}^\lambda$ par Π^λ.

La projection $(f_\lambda)_{\lambda \in \gamma_1^+}$ dans la partie continue d'une fonction $f \in \mathcal{C}_c^\infty(G/K)$ peut être précisée, elle est donnée par $f_\lambda = \pi^\lambda(f)(\psi_\lambda)$.

Enfin, pour la décomposition spectrale de $L^2(G/I)$, qui s'identifie à l'espace $L^2(\widehat{A})$ des fonctions de carré sommable sur les arêtes de l'arbre, on introduit la fonction $\iota_\lambda \in \mathcal{H}^\lambda$, de support IB, et dont la restriction à I est constante et vaut $(\mathrm{mes}(I))^{-1} = \frac{1}{q+1}$.

Théorème 9. *La représentation régulière de G dans $L^2(G/I)$ est isomorphe à la somme directe d'une partie discrète constituée de la somme directe des deux représentations T^0 et $\mathbf{St}$, et d'une partie continue qui est l'intégrale hilbertienne des $\mathcal{H}^\lambda$ pour la mesure dm de support γ_1^+.*

$$\int_{\gamma_1^+}^{\oplus} \mathcal{H}^\lambda dm(\lambda) \oplus T^0 \oplus \mathbf{St},$$

où G opère dans $\mathcal{H}^\lambda$ par Π^λ.

Le plongement f_λ dans la partie continue de la décomposition spectrale est donné pour $f \in \mathcal{C}_c^\infty(G/I)$ par

$$f_\lambda = \pi^\lambda(f)(\iota_\lambda).$$

Bibliographie

[Ca-1] P. CARTIER: Géométrie et analyse sur les arbres, *Séminaire Bourbaki, 1971–72 exposé 407. Lectures Notes in Math. Springer-Verlag 1973, p.123–140.*

[Ca-2] P. CARTIER: Fonctions harmoniques sur un arbre, *Symposia Mathematica 9, 1972, p.203–270.*

[Ch-1] F. CHOUCROUN: Analyse harmonique des groupes d'automorphismes d'arbres de Bruhat-Tits, *Mémoire n°58, 1994, supplément au Bull. S.M.F.122(3).*

[Ch-2] F. CHOUCROUN: Arbres, espaces ultramétriques et bases de structure uniforme *Geometriae Dedicata 53 (1994) p.69–74.*

[L] S. LANG: $SL_2(\mathbf{R})$ *Addison-Wesley (1975).*

[S] J.P. SERRE: Arbres, amalgames, SL_2 *Astérisque, 46 (1977).*

Francis M. Choucroun
F. C. Mathématiques, Bât 425
Université Paris Sud
91405 Orsay-Cedex

choucrou@math.u-psud.fr

Progress in Probability, Vol. 40
© 1996 Birkhäuser Verlag Basel/Switzerland

Trees and Non-Archimedean Topologies

G. CHRISTOL

Mathematics Subject Classification: 82D30

Keywords: non-archimedean spaces.

Abstract. The aim of this text is to give a non-Archimedean interpretation of trees. This point of view yields natural (and useful!) extensions of the notion of tree.

Resumé. Les liens entre arbre et espaces ultramétriques compacts ou localement compacts permettent de proposer plusieurs généralisations de la notion d'arbre. L'une d'elle est implicitement utilisée dans le truc des répliques pour l'étude théorique des verres de spin.

1   From trees to non-Archimedean spaces

Let T be an infinite tree with no finite branch. Given some vertex O, let $F(T)$ denote the set of infinite paths in T starting from O and with no back-tracking. It is easy to see that $F(T)$ is independent of the given origin O the rule being to identify two paths with an infinite number of common vertices. We will call $F(T)$ the *set of leaves* of T and, for some vertex O of T, we denote by $\{O \to x\}$ the infinite set of vertices that lie in the path defining x.

The set $F(T)$ is endowed with natural non-Archimedean distances which are built in various ways but give all the same topology.

1.1 Compact distances

Choose a vertex O of T and a sequence $\{d_n\}_{n\in\mathbb{N}}$ of (positive) real numbers that decreases to 0.

For x and y in $F(T)$ with $x \neq y$ the paths $\{O \to x\}$ and $\{O \to y\}$ have only a finite number of common vertices. Define

$$d(x,y) = 0 \quad \text{if } x = y,$$
$$d(x,y) = d_n \quad \text{if } x \neq y \text{ and } \{O \to x\} \cap \{O \to y\} = \{O, s_1, \ldots, s_n\}.$$

One verifies immediately that d is a non-Archimedean distance (namely it satisfies the non-Archimedean triangle inequality $d(x,z) \leq \max\{d(x,y), d(y,z)\}$). Moreover, it is not difficult to prove that the induced topology on $F(T)$ is independent of the choices we made (for O and $\{d_n\}$) and makes $F(T)$ a compact space.

1.2 Locally compact distances

Choose a leaf ∞ in $F(T)$, a vertex O in T and a decreasing map $\{d_n\}_{n\in\mathbb{Z}}$ from $\mathbb{Z}$ to $\mathbb{R}$ such that $\lim_{n\to+\infty} d_n = 0$.

For each leaf x such that $\{O \to x\} \cap \{O \to \infty\} = \{O\}$ there is a unique one to one application i_x from $\mathbb{Z}$ to $\{\infty \to x\} = \{O \to x\} \cup \{O \to \infty\}$ such that

$$i_x(0) = O, \qquad \{O \to x\} = i_x(\mathbb{N}),$$

there is an edge connecting $i_x(n)$ and $i_x(n+1)$.

More generally, for any leaf $x \neq \infty$, there exists an integer m such that $\{O, x\} \cap \{O \to \infty\} = \{O, s_1, \ldots, s_m\}$, whence it is easy to construct a (unique) one to one application i from $\mathbb{Z}$ to $\{x, \infty\} = \{O \to x\} \cup \{O \to \infty\} - \{O, s_1, \ldots, s_{m-1}\}$ such that

$$i_x(-m) = s_m, \qquad \{s_m \to x\} = i_x(\mathbb{Z} \cap [-m, \infty[),$$

there is an edge connecting $i_x(n)$ and $i_x(n+1)$.

Now, by construction, for x and y in $F(T) - \{\infty\}$, $x \neq y$, the paths $\{\infty \to x\}$ and $\{\infty \to y\}$ have same beginning on which the applications i_x and i_y are the same. Define

$$d(x, y) = 0 \quad \text{if } x = y$$
$$d(x, y) = d_n \quad \text{if } x \neq y \text{ and } n = \max\{m \in \mathbb{Z} \; ; i_x(m) = i_y(m)\}.$$

One verifies immediately that d is a non-Archimedean distance on $F(T) - \{\infty\}$, that the induced topology is independent of the choices for O and $\{d_n\}$ and makes $F(T)$ a locally compact space. This locally compact topology on $F(T) - \{\infty\}$ is nothing else than the one induced by the compact topology on $F(T)$ as defined in 1.1.

2 From non-Archimedean spaces to trees

Let F be an infinite non-Archimedean compact space (namely a set F endowed with a non-Archimedean distance d and which is compact for the induced topology).

Let $D(a, r)$ denote the disk centered in a with radius r, namely

$$D(a, r) = \{x \in F; d(x, a) \leq r\}.$$

As the distance d is a continuous function from the compact space $F \times F$ to $\mathbb{R}$, its image $\mathcal{I}$ is compact. Moreover, if $\{r_n = d(a_n, b_n)\}$ is an infinite sequence of distinct elements in $\mathcal{I}$, one can assume, by compactness, that the sequences a_n and b_n converge, namely $a_n \to a$ and $b_n \to b$. If $a \neq b$, then by the non-Archimedean property, one would have $r_n = d(a_n, b_n) = d(a, b)$ for n large enough, a contradiction. Hence, F being infinite and Hausdorff, the only possibility is to have

$$\mathcal{I} = \{d(x, y); x, y \in F\} = \{d_n\}_{n \in \mathbb{N}} \cup \{0\}$$

where $\{d_n\}$ is a sequence that decreases to 0.

Let T be the graph whose vertices are disks in F (with radius d_n for some n) and whose edges are couples of concentric disks (i.e. not disjoint) with respective radius d_n and d_{n+1}.

Proposition: *The graph T is a tree whose set of leaves is F.*

The proof is easy when noticing that each disk $D(a, d_n)$ is a finite (by compactness) union of disjoint (by non-Achimedeanness) disks $D(b_i, d_{n+1})$.

Remark: One has a realization for T. On $F \times \mathbb{R}$, define the equivalence

$$(x, r) \sim (y, s) \text{ iff } (r = s \text{ and } d(x, y) \leq r \text{ (i.e. } D(x, r) = D(y, r))) \,.$$

Then, for $r \in \{d_n\}$, each equivalence class $\overline{(x, r)}$ realizes one vertex of T and, for $d_n < r < d_{n+1}$, the equivalence classes $\overline{(x, r)}$ realize the edges of T that connect $\overline{(x, d_n)}$ and $\overline{(x, d_{n+1})}$.

3 Inverse limit: construction of $\widehat{\mathbb{Z}}$

By means of inverse limit, one can construct simultaneously the set $F(T)$ and its topology. More precisely, choose a vertex O in T and denote by S_n the set of vertices of T connected to O by a path with exactly n edges. For S in S_{n+1} define $p(S)$ to be the only vertex in S_n such that $(S, p(S))$ is an edge. For T, to have no finite branch, means that p is onto for all n. Then by definition

$$F(T) = \varprojlim_{n \in \mathbb{N}} T_n.$$

Namely, $F(T)$ is the unique (up to isomorphism) smallest set with the coarsest topology such that there exist continuous applications $p_n : F(T) \to T_n$ (for T_n endowed with discrete topology) such that $p \circ p_{n+1} = p_n$.

Actually, pointed trees and projective systems of finite sets indexed by integers are the same. For instance the set $\mathbb{Z}_p$ of p-adic integers is defined by

$$\mathbb{Z}_p = \varprojlim_{n \in \mathbb{N}} \mathbb{Z}/p^n\mathbb{Z} \,.$$

To generalize the notion of tree, one can replace the index set $\mathbb{N}$ by a more general one. It is enough to work with a "directed set" (namely a partially ordered set I such that for each pair i, j in I there exists k in I such that i and j are both smaller than k). Denote by $\mathbb{N}^{\times}$ the set $\mathbb{N}^*$ endowed with the division partial ordering (i smaller than j means i divides j) and use the "canonical projections"

$$\mathbb{Z}/nm\mathbb{Z} \xrightarrow{\; p_{nm,m} \;} \mathbb{Z}/n\mathbb{Z}$$

defined by reduction modulo n. The inverse limit of the system $\{\mathbb{Z}/n\mathbb{Z}, p_{n,m}\}$ is a topological set denoted by $\widehat{\mathbb{Z}}$. It is easily seen that

$$\widehat{\mathbb{Z}} = \varprojlim_{n \in \mathbb{N}^{\times}} \mathbb{Z}/n\mathbb{Z} = \prod_{p \text{ prime}} \mathbb{Z}_p \,. \tag{1}$$

Actually, $\widehat{\mathbb{Z}}$ is a metric space and one can endow it with many equivalent distances. We will describe some of them.

Let $\{u_n\}$ be a sequence of integers. We will write $u_n \times \to \infty$ and say that the sequence $\{u_n\}$ *goes multiplicatively to infinity* if the following two conditions are satisfied:

1) $u_0 = 1$ and, for all n, u_{n+1} is a multiple of u_n.
2) $(\forall m \in \mathbb{N})(\exists N \in \mathbb{N}) : u_n$ is multiple of m for $n \geq N$.

Under these circumstances, the canonical projections $\mathbb{Z}/u_{n+1} \to \mathbb{Z}/u_n\mathbb{Z}$ form a classical inverse system indexed by $\mathbb{N}$ and one easily verifies

$$\widehat{\mathbb{Z}} = \varprojlim_{n \in \mathbb{N}} \mathbb{Z}/u_n\mathbb{Z} . \tag{2}$$

Let $\varphi_n : \widehat{\mathbb{Z}} \to \mathbb{Z}/u_n\mathbb{Z}$ be the canonical projection. For any sequence d_n of real numbers that decreases to 0, the relation

$$d(a,b) = d_n \qquad \text{with} \quad n = \max\{k; \varphi_k(a) = \varphi_k(b)\}$$

defines a distance d on $\widehat{\mathbb{Z}}$ associated with the topology given by (2) which is also given by (1).

One can describe even more concretely the distance d by building step by step the (infinite) distance table using the following process:

a) D_0 is the 1×1 matrix $[0]$.
b) For $n \geq 0$, D_{n+1} is the $u_{n+1} \times u_{n+1}$ matrix built up from D_n by substituting the entry $D_{n,i,j}$, lying on i-th line and j-th column, by the square block of dimension (u_{n+1}/u_n) given by the rules:

    b1) for $i \neq j$, all entries are equal to $D_{n,i,j}$,
    b2) for $i = j$, the diagonal entries are 0 and all others are equal to d_n.

Rules b1) and b2) are straightforward consequences of the non-Archimedean property. As an example, we give the matrix D_3 associated to sequences $u_1 = 2$, $u_2 = 6$, $u_3 = 12$, $d_0 = 1$, $d_1 = \alpha$, $d_2 = \beta$. This is nothing other than the distance table over $\mathbb{Z}/12\,\mathbb{Z}$ when numbers are lexicographically ordered by their residues modulo 2, 6 and 12.

One recognizes, up to slight modifications, Parisi's matrices used in the replica trick in spin glasses theory (see [3]). The next section will be devoted to an interpretation of the replica trick calculus.

Remark: Let $\{1, u_1, \ldots, u_n, \ldots\}$ be a sequence that goes multiplicatively to infinity and let T_u be the pointed tree in which each branch of the n-th level is ramified in u_{n+1}/u_n branches of the $(n+1)$-th level. Actually T_u has exactly u_n branches of n-th level. Relation (2) shows that $\widehat{\mathbb{Z}} = F(T_u)$. Now every pointed tree is a subtree of some T_u. In that respect, one can see $\widehat{\widehat{\mathbb{Z}}}$ as a generic tree for everything concerning sets of leaves.

	11	5	9	3	7	1	10	4	8	2	6	0
11	0	β	α	α	α	α	1	1	1	1	1	1
5	β	0	α	α	α	α	1	1	1	1	1	1
9	α	α	0	β	α	α	1	1	1	1	1	1
3	α	α	β	0	α	α	1	1	1	1	1	1
7	α	α	α	α	0	β	1	1	1	1	1	1
1	α	α	α	α	β	0	1	1	1	1	1	1
10	1	1	1	1	1	1	0	β	α	α	α	α
4	1	1	1	1	1	1	β	0	α	α	α	α
8	1	1	1	1	1	1	α	α	0	β	α	α
2	1	1	1	1	1	1	α	α	β	0	α	α
6	1	1	1	1	1	1	α	α	α	α	0	β
0	1	1	1	1	1	1	α	α	α	α	β	0

4   The replica trick

Based on the following remark:

$$\log(z) = \lim_{s \to 0} \frac{z^s - 1}{s},$$

the trick of replica calculus is to consider a sequence of integers that goes to zero. This is very uncomfortable when working in $\mathbb{R}$ but natural when working in $\widehat{\mathbb{Z}}$. Namely, the sequence u_n goes to 0 in $\widehat{\mathbb{Z}}$ iff it becomes a multiple of every fixed integer. This is almost the definition of a sequence that goes multiplicatively to infinity.

4.1 First step and integration over $\widehat{\mathbb{Z}}$

By means of integration on $\widehat{\mathbb{Z}}$, one can recover, at least formally, the main formulae of the replica trick calculus.

For any topological field $\mathbb{A}$, an $\mathbb{A}$-valued measure μ on $\widehat{\mathbb{Z}}$ can be defined by a system $\{\mu_n\}_{n \in \mathbb{N}^\times}$ of applications from $\mathbb{Z}/n\mathbb{Z}$ into $\mathbb{A}$ such that

$$\mu_n(a) = \sum_{p_{nm,m}(b)=a} \mu_{nm}(b)$$

for every pair (n, m) and $a \in \mathbb{Z}/n\mathbb{Z}$. To avoid difficulties, it is better to limit oneself to bounded measures, i.e. 0 is not a limit point of the set $\{1/\mu_n(a)\}_{n \in \mathbb{N}, a \in \mathbb{Z}/n\mathbb{Z}}$.

For $g : \widehat{\mathbb{Z}} \to \mathbb{A}$, one sets

$$\int_{\widehat{\mathbb{Z}}} g(x) \, d\mu(x) = \lim_{n \times \to \infty} \sum_{a=0}^{n-1} g(a) \, \mu_n(a)$$

and one verifies that the limit exists for bounded measures and continuous g.

For instance, as a compact group, $\widehat{\mathbb{Z}}$ has a Haar measure, obtained as the inverse limit of Haar measures η_n on $\mathbb{Z}/n\mathbb{Z}$, namely the equidistributed ones:

$$\eta_n(a) = \frac{1}{n}.$$

Notice that, for $\mathbb{A}$ containing some $\mathbb{Q}_p$, the Haar measure is not bounded (recall that being near zero in $\mathbb{Q}_p$ means to be highly divisible by p).

For a continuous application $g : \widehat{\mathbb{Z}} \to \mathbb{C}$, and for any sequence $\{u_n\}$ that goes multiplicatively to infinity, one has

$$\int_{\widehat{\mathbb{Z}}} g(x)\, d\eta(x) = \lim_{n \times \to \infty} \frac{1}{n} \sum_{a=0}^{n-1} g(a) = \lim_{n \to \infty} \frac{1}{u_n} \sum_{a=0}^{u_n-1} g(a).$$

Now, given an application $f : \mathbb{R} \to \mathbb{C}$, one obtains:

$$\int_{\widehat{\mathbb{Z}}} f(d(x,0))\, d\eta(x) = \lim_{n \to \infty} \frac{1}{u_n} \sum_{a=0}^{u_n-1} f(d(a,0)) \tag{3}$$

$$= \lim_{n \to \infty} \frac{1}{u_n - 1} \sum_{a=1}^{u_n-1} f(d(a,0)) \tag{4}$$

$$= \lim_{n \to \infty} \frac{1}{u_n(u_n - 1)} \sum_{b=0}^{u_n-1} \sum_{\substack{a=0 \\ a \neq b}}^{u_n-1} f(d(a,b))$$

$$= \lim_{n \to \infty} \frac{1}{u_n(u_n - 1)} \sum_{\substack{i,j \\ i \neq j}} f(D_{n,i,j}).$$

As u_n goes to infinity, $\lim f(0)/u_n = 0$. In this step, between (3) and (4), one changes u_n to $(u_n - 1)$ only for the sake of elegance.

On the other hand, the repartition function

$$x(q) = \eta\{z \in \widehat{\mathbb{Z}} \; ; d(z,0) < q\} = \frac{1}{u_n} \quad \text{for } q \in]d_n, d_{n-1}]$$

has almost an inverse, namely

$$q(x) = d_n \quad \text{for } x \in \left[\frac{1}{u_{n+1}}, \frac{1}{u_n} \right[,$$

and one finds

$$\int_{\widehat{\mathbb{Z}}} f(d(x,0))\, d\eta(x) = \int_0^1 f(q(x))\, dx = f(1) + \sum_{k=1}^{\infty} \frac{1}{u_k}\left(f(d_k) - f(d_{k-1})\right).$$

All these formulas remind us of the ones used in the replica trick, but there is a big gap between them. Here, when u_n appears outside the functions, it has to be considered in $\mathbb{R}$ and then goes to infinity. In the replica trick, on the contrary, one considers u_n going to 0. In particular, the sequences u_n and $(u_n - 1)$ have quite distinct behaviors and changing one to the other is no longer harmless.

4.2 Second step: actual computation

The replica trick actually uses the following "philosophy":

"By construction, $u_n \geq u_{n-1} \geq \cdots \geq u_1 \geq 1$. In the $u_n \to 0$ limit this becomes $1 \geq u_1 \geq \cdots \geq u_n$ and in the limit $n \to \infty$, u_k becomes a continuous variable x, $0 \leq x \leq 1$, and $u_{k+1} \to x + dx$."

To give meaning to this, the first idea is to consider $\widehat{\mathbb{Z}}$-valued measures. Then the sequence $\{u_n\}$ actually goes to zero. But this does not seem to give a convincing explanation.

It seems better to consider the replica trick as integration over an unknown, and perhaps non compact, non-Archimedean space endowed with a measure μ.

Actually, using formula (4) and $\lim u_n = 0$, one gets

$$\lim_{n\to\infty} \frac{1}{u_n - 1} \sum_{a=0}^{u_n-1} f(d(a,0)) = -\lim_{n\to\infty} \sum_{a=0}^{u_n-1} f(d(a,0))$$

$$= -\lim_{n\to\infty} u_n \sum_{k=0}^{n} \left(\frac{1}{u_k} - \frac{1}{u_{k+1}} \right) f(d_k).$$

Notice that u_k/u_n is the measure of the disk $D(0, d_k)$ in $\widehat{\mathbb{Z}}$, so, setting $m(x) = \mu(D(0,x))$, one obtains

$$u_n \sum_{k=0}^{n} \left(\frac{1}{u_k} - \frac{1}{u_{k+1}} \right) f(d_k) = \sum_{k=0}^{n} \frac{m(d_{k+1}) - m(d_k)}{m(d_{k+1})\, m(d_k)} f(d_k).$$

Using replica philosophy, one has $d_k \to x$, $d_{k+1} \to x + dx$ and $n \to \infty$. In the limit, we then get

$$\int_0^\infty f(x) \frac{-dm(x)}{m(x)^2} \; .$$

In fact, in the spin glasses model, one replica is not enough, only the relative situations of r replica ($r \geq 2$) makes sense (in random limit). For that purpose, one has to integrate over $\widehat{\mathbb{Z}}^r$ instead of $\widehat{\mathbb{Z}}$. Applying the preceding method leads to results analogous (but a little bit different) to those obtained by the replica method.

5 Quasi-polyhedrons

It is a natural idea to consider more general non-Archimedean spaces. We will only speak about the one dimensional case. The multi-dimensional case leads to the so called "Bruhat-Tits buildings".

Let $\mathbb{K}$ be a field which is complete for a non archimedean valuation. An affine space F over $\mathbb{K}$ is given by a commutative Banach $\mathbb{K}$-algebra $\mathcal{A}$ (roughly speaking it is the algebra of "holomorphic" functions on F). The "closed" points of F are the maximal ideals of $\mathcal{A}$. Actually, "points of F" are prime ideals of $\mathcal{A}$ but in the one dimensional case prime ideals are maximal.

We think of F as the leaves of some generalized tree and follow Berkovitch [1] to construct the generalized tree itself or, more precisely, its realization $\mathcal{T}$. A point in $\mathcal{T}$ is a multiplicative semi-norm on $\mathcal{A}$, namely a continuous application $f \mapsto \|f\|$ from $\mathcal{A}$ to $\mathbb{R}^+$, such that

$$\|fg\| = \|f\|\,\|g\| \text{ and } \|f + g\| \leq \max(\|f\|, \|g\|).$$

Points of $\mathcal{A}$ are points of $\mathcal{T}$: to a prime ideal I of $\mathcal{A}$ is associated the semi-norm

$$\|f\| = \begin{cases} 0 & \text{if } f \in I \\ 1 & \text{if } f \notin I \end{cases}.$$

One endows $\mathcal{T}$ with the coarsest topology such that, for each f in $\mathcal{A}$, the application $\|.\| \mapsto \|f\|$ is continuous.

Example: The affine line $\mathbb{A}$ over $\mathbb{K}$ is associated to the $\mathbb{K}$-algebra $\mathbb{K}[x]$. Actually, $\mathbb{K}[x]$ is not a Banach algebra and one has to "cover" $\mathbb{A}$ with affinoids, but we will skip this technical difficulty. Beside points of F, $\mathcal{T}$ has mainly two kinds of points:

1) "closed" disks of $\mathbb{K}$: to a disk $D = D(a, r)$ with $r = |b|$ for some b in $\mathbb{K}$, one associates the semi-norm $\|f\| = \max_{x \in D} |f(x)|$. These points are considered as vertices of T.

2) "other disks": to a disk $D = D(a, r)$ where r is not the absolute value of some element in $\mathbb{K}$, one associates the semi-norm $\|\sum a_i\, x^i\| = \max |a_i|\, r^i$. These points form the edges of $\mathcal{T}$.

There can also appear rather pathological points corresponding to empty intersections of decreasing sequences of disks. In the case of a locally compact field one just recovers the tree described in 2.

In the general one-dimensional case $\mathcal{T}$ is a "simply connected quasi-polyhedron", i.e. a topological space with the following properties:

i) For any x, y in $\mathcal{T}$ there exists a unique closed subset $\ell_{x,y}$ in $\mathcal{T}$ homeomorphic to $[0, 1]$ with end points x and y.

ii) $\mathcal{T}$ is locally compact. A basis of the topology is formed by open U such that $\overline{U} - U$ is finite.

iii) All connected open subsets of $\mathcal{T}$ are countable at infinity.

This is a nice generalization of trees. Roughly speaking, it is an "hairy" tree in which each point of an edge can be a vertex and each vertex can be on infinitely many edges.

Bibliography

[1] BERKOVICH V.G.: Spectral theory and analytic geometry over non Archimedean fields, *Mathematical Surveys and Monographs* **33**, Providence (1990).

[2] CHOUCROUN F.: Arbres, espaces ultramétriques et bases de structure uniforme, *Geometriae Dedicata* **53** (1994) 69-74.

[3] MÉZARD M., PARISI G., SOURLAS N., TOULOUSE G., VIRASORO M.: Replica symmetry breaking and the nature of the spin glass phase, *Phys. Rev. Lett.* **55** (1984) 1156.

Gilles Christol
Université de Paris VI
Mathématique case 247
4, Place Jussieu
75250 Paris Cedex 05, France

christol@mathp6.jussieu.fr

4. Large Deviations

Progress in Probability, Vol. 40
© 1996 Birkhäuser Verlag Basel/Switzerland

Arbres et Grandes Déviations

ALBERT BENASSI

Mathematics Subject Classification: 05C05, 60F10, 60G18

Keywords: Trees, large deviations, entropy

Abstract. On the vertices of a tree A we dispose real and iid random variables. We study the maximum S_n^* of the sums of those variables encountered on the length n branches of A. We introduce the entropic number $e(A)$ of A. If Λ denotes the large deviations function of the basic law, we show that $\limsup_{n\to\infty} \dfrac{S_n^*}{n} \leq \Lambda^{-1}[\log(e(A))]$, with equality in the gaussian case.

Résumé. Sur les sommets d'un arbre A sont disposées des variables aléatoires réelles indépendantes et de même loi. Nous étudions les maxima S_n^* des sommes de ces variables rencontrées sur les branches de longueur n de A. Nous définissons le taux entropique $e(A)$ de A, et si Λ désigne la fonction de grandes déviations de la variable aléatoire de base, nous prouvons que $\limsup_{n\to\infty} \dfrac{S_n^*}{n} \leq \Lambda^{-1}[\log(e(A))]$, avec égalité dans le cas gaussien.

1 Introduction

1.1 Soit A un arbre enraciné en O, localement fini. Supposons que tous les sommets de A ont pour degré au moins deux. Si $\tau \in A$, notons C_τ le plus court chemin entre O et τ dans A. Le nombre d'arêtes de C_τ est noté $|\tau|$. Si $\delta \in A$, écrivons $\delta \leq \tau$ quand $\delta \in C_\tau$; de même notons $C_{\delta,\tau}$ l'ensemble $\{t \in A \mid \delta \leq t \leq \tau\}$, le chemin de δ à τ qui a pour longueur $|\tau| - |\delta|$. Dans le même esprit, l'ensemble $\theta = \{t \in A \mid \tau \leq t \ ; \ |t| - |\tau| = h(\theta)\}$ est le sous-arbre de A de racine $\tau(\theta)$ et de hauteur $h(\theta)$; il a pour frontière $\Sigma(\theta) = \{t \in \theta \mid |t| - |\tau| = h(\theta)\}$. Nous noterons par $|\theta|$ le cardinal de $\Sigma(\theta)$. Deux sous-arbres θ_1 et θ_2 de A vérifient $\theta_2 \subset \theta_1$ si $\tau(\theta_1) \leq \tau(\theta_2)$. Notons A_n le sous-arbre de A de racine O et de hauteur n.

Un certain nombre de caractéristiques géométriques sont attachées à un arbre A. Notamment sa dimension (de Minkowski cf [BP]), $\log(d(A)) = \limsup_{n\to\infty} \dfrac{1}{n}\log(|A_n|)$, et son taux de branchement $b(A)$ caractérisant la "ramification apparente" dans A (cf [L]).

$$b(A) = inf\{\lambda > 0 : \liminf_{\pi\to\infty} \sum_{\sigma\in\pi} \lambda^{-|\sigma|} = 0\}$$

où π est un sous-ensemble de sommets de A tel que toute géodésique de A partant de la racine 0 ne coupe π qu'en un seul point. Le symbole $\pi \to \infty$ signifie que le (ou l'un des) point de π le plus proche de 0 s'en éloigne indéfiniment. Le fait que $1 < b(A) < d(A)$ signifie que A contient un sous-arbre enraciné en 0 qui est quasi-sphérique (c'est-à-dire dont le taux de branchement est égal à son taux de croissance : $b(A) = d(A)$).

Parmi les problèmes probabilistes que l'on peut étudier sur A se pose celui des grandes déviations de sommes de variables aléatoires réelles (v.a.r.) dont voici l'énoncé.

1.2 Grandes déviations

Soit X une v.a.r. et soit $\{X_t; t \in A\}$ une famille iid de v.a.r. de loi commune celle de X. Posons $S(t) = \sum_{\tau \in C_t} X_\tau$. Nous voulons étudier la quantité $U(X, A)$ définie par

$$(1) \qquad U(X, A) = \limsup_{n \to \infty} \frac{S^*(A_n)}{n},$$

avec, si θ est un sous-arbre de A

$$S^*(\theta) = max\Big\{ \sum_{\tau(\theta) \leq \tau \leq t} X_\tau, \quad t \in \Sigma(\theta) \Big\}.$$

Dans cette note, nous ferons l'hypothèse $\underline{H}$ suivante :

$$\underline{H} \qquad\qquad (i) \ \ X \text{ possède tous ses moments exponentiels,}$$
$$(ii) \ \ d(A) < \infty.$$

Notons que sous $\underline{H}$, $U(X, A) < \infty$ et que $U(X, A)$ est, avec probabilité un, une constante. Introduisons la fonction des grandes déviations Λ de X par

$$\lambda(u) = \log[E(e^{uX})], u \in \mathbb{R} \ \ ; \ \ \Lambda(v) = max\{uv - \lambda(u), u \in \mathbb{R}\}.$$

Lemme 1. [B,P]

(i) $b(A) = d(A) \Longrightarrow U(X, A) = \Lambda^{-1}[\log(d(A))],$

(ii) *dans le cas général,* $\Lambda^{-1}[\log(d(A))] \geq U(X, A) \geq \Lambda^{-1}[\log \sqrt{d(A)}]$

En général, ces inégalités sont strictes et ne sont pas optimales. Le but de cette note est de donner des caractéristiques géométriques de A permettant de les remplacer par des inégalités optimales.

2   Modes d'un arbre

Soit T une famille infinie de sous-arbres de A de hauteur finie. Donnons la description de ces objets. Soit $\theta \in A$, notons $n(\theta)$ le nombre défini par $n(\theta) = inf\{n \mid \theta \subset A_n\}$. $n(\theta)$ est la distance de 0 à la racine $\tau(\theta)$ de θ. Introduisons le taux de croissance $g(T)$ et le taux d'explosion $q(T)$ respectivement par :

$$\log[g(T)] = \limsup_{\theta \in T, \theta \to \infty} \left(\frac{\log(|\theta|)}{n(\theta)} \right)$$

$$\log[q(T)] = \limsup_{\theta \in T, \theta \to \infty} \left(\frac{\log(|\theta|)}{h(\theta)} \right)$$

ainsi que la hauteur relative $r(T)$ et que la corrélation $C(T)$ par

$$r(T) = \limsup_{\theta \in T, \theta \to \infty} \left(\frac{h(\theta)}{n(\theta)} \right) \; ;$$

si

$$C(\theta) = \frac{2}{|\theta|(|\theta| - 1)} \sum_{\substack{t,s \in \Sigma(\theta) \\ t \neq s}} (t,s)_\theta$$

avec $(t,s)_\theta$ le nombre d'arêtes communes à t et s dans θ,

$$C(T) = \limsup_{\theta \in T, \theta \to \infty} \big[C(\theta) \big].$$

Nous avons convenu que $\theta \to \infty \Leftrightarrow h(\theta) \to \infty$ et $|\theta| \to \infty$. La famille $T = (A_n; n \in \mathbb{N})$ est un mode, le mode trivial.

Soient T_1 et T_2 deux familles de sous-arbres de A, nous dirons que $T_1 \subset T_2$ si $\forall \theta_1 \in T_1, \exists \theta_2 \in T_2$ et $\theta_1 \subset \theta_2$.

Mode. Une famille $T(r,q)$ est un mode de A si
1) $r(T) > 0, q(T) < \infty, C(T) < \infty$, et les limites supérieures les définissant sont des limites.
2) T est une famille ordonnée.
3) Si $T'(r',q')$ vérifie (1) et (2) et est telle que $T(r,q) \subset T'(r',q')$ alors $r = r', q = q'$ et $C(T) \leq C(T')$.

Projection d'un mode. Si T est un mode de A, T_k désigne la famille finie des $\theta \cap A_k := \theta_k$, $k \in \mathbb{N}$, $\theta \in T$. Une suite $(k_n; n \geq 0)$ de $\mathbb{N}$ est admissible pour T si $\theta_{k_n} = \theta_{k_{n(\theta)}}$, $\forall n \in \mathbb{N}$.

Lemme 2. *Sous $\underline{H}(ii)$, $\forall \varepsilon > 0, \exists N_\varepsilon \in \mathbb{N}$ et des modes $T(r_i, q_i)$, $i = 1, \ldots, N_\varepsilon$ tels que*

$$\limsup_{n \to \infty} \left[\max_{\theta_i \in T(r_i, q_i) \; i=1,\ldots,N_\varepsilon} \frac{1}{n} \log \left[\sum_{i=1}^{N_\varepsilon} | (\theta_i)_n | \right] \right] \geq d(A) - \varepsilon \; .$$

Preuve. 1) Dans le cas quasi-sphérique $T(1, d(A))$ est le mode trivial.
2) Dans le cas général $(b(A) < d(A))$ il existe au moins un mode, si non il serait impossible de générer de l'ordre de $d(A)^n$ (en convenant que : si r_n et l_n sont deux suites positives, on écrira $r_n \overset{\log}{\approx} l_n$ si $\frac{\log(r_n)}{n} \approx \frac{\log(l_n)}{n}$) branches de A_n.
3) Dans ce cas, les modes apparaissent successivement dans A_n. Le lemme est donc un résultat de continuité. $\qquad \square$

3   Grandes déviations

Commençons par étudier les grandes déviations sur les modes. Soit $T(r, q)$ un mode de A.

Lemme 3. *Si A et X vérifient $\underline{H}$, alors*

$$(3) \qquad \limsup_{n \to \infty} \frac{S^*(T_n(r, q))}{n} = r\Lambda^{-1}(\log q) \quad pps.$$

Preuve. On peut, en adaptant la définition du nombre de branchement d'un arbre, définir le nombre de branchement $b(T)$ d'un mode $T(r, q)$ de A. Mais alors $T(r, q)$ est quasi-sphérique en ce sens que $b(T) = q(T)$. Dans ce cas le lemme 1, partie (i) permet de conclure que

$$\limsup_{n \to \infty} \frac{S^*(\theta_n(T))}{rn} = \Lambda^{-1}(\log q(t)),$$

d'où le lemme. $\qquad\qquad\qquad\qquad\qquad\qquad\qquad\qquad\qquad\qquad\qquad\qquad\qquad$ $\square$

Voici le résultat principal de cette note.

Théorème 1 *Sous $\underline{H}$, il existe un mode $T(r(X), q(X))$ de A tel que*

$$U(X, A) = r(X)\Lambda^{-1}[\log(q(X)].$$

Preuve. Soit T un mode de A, notons $U(X, T)$ la quantité (3) du lemme 3. Classons les modes de A ; soit $\delta > 0$, Cl_i désigne

$$Cl_i = \{T \text{ modes de } A \mid U(X, A)\delta i \leq U(X, T) \leq U(X, A)\delta(i+1) ; 1 \leq i \leq \left[\frac{1}{\delta}\right]+1\}$$

Posons alors $(Cl_i)^n = \{T_n , T \in Cl_i\}$, et $S_n^*(Cl_i) = \max_{T \in Cl_i} S_n^*(T)$. Grace au lemme 2, $\forall \varepsilon > 0, \exists N_\varepsilon(X)$ et des modes $T_1(X), \ldots, T_{N_\varepsilon}(X)$ satisfaisant (3). Notons A_n^ε le sous-arbre de A_n engendré par $T_1(X), \ldots, T_{N_\varepsilon}(X)$ et R_n^ε celui engendré par $A - A_n^\varepsilon$. Par continuité, il vient

$$\limsup_{n \to \infty} \frac{S^*(R_n^\varepsilon)}{n} < U(X, A).$$

Le lemme 3 permet alors de conclure, en choisissant bien T_i dans Cl_i avec $\delta = \frac{1}{N_\varepsilon(X)}$. $\qquad\qquad\qquad\qquad\qquad\qquad\qquad\qquad\qquad\qquad\qquad\qquad\qquad$ $\square$

La loi gaussienne joue un rôle particulier. Le théorème 1 permet d'associer à la loi $N(0, 1)$ un mode $T(r(G), q(G))$ qui porte la grande déviation des sommes gaussiennes. Définissons le taux entropique $e(A)$ de A par $e(A) = q(G)^{r^2(G)}$.

Lemme 4.

$$e(A) = d(G)^{r(G)}$$

Preuve. Nous avons nécessairement

$$q(G)^{r(G)n} \overset{\log}{\approx} d(A)^n \text{ (sinon } U(G, A) > U(G, T(G))),$$

d'où le lemme. $\qquad\qquad\qquad\qquad\qquad\qquad\qquad\qquad\qquad\qquad\qquad\qquad\qquad$ $\square$

Finalement nous avons

Corollaire. $\forall X$ vérifiant $\underline{H}$ (i),

$$r(G)\Lambda^{-1}\big(\log\ q(G)\big) \leq U(X,A) \leq \Lambda^{-1}\big(\log\ e(A)\big).$$

4   Exemples

Soit X la loi : $P(X=1) = P(X=-1) = 1/2$ et G la loi $N(0,1)$.

Exemple 1

On construit l'arbre A de la façon suivante ; - De la racine partent 3 branches.

- A l'étape n on partage $\Sigma(A_n)$ en trois parties égales numérotées de 1 à 3. Les éléments 1 et 2 ont deux descendants, ceux du numéro 3 ont 5 descendants. Dans ce cas

$$b(A) = 2,\quad d(A) = 3,\quad q(A) = 5,\quad r(A) = \frac{\log\ 3}{\log\ 5}.$$

X sélectionne le mode $T(1,2)$, G sélectionne le mode $T(r(A),q(A))$, $e(A) = 3^{\frac{\log\ 3}{\log\ 5}}$.

Exemple 2

On construit l'arbre A de la façon suivante ; - De la racine partent 5 branches.

- A l'étape n on partage $\Sigma(A_n)$ en cinq parties égales numérotées de 1 à 5. Les éléments de numéro 1, 2 et 3 ont un descendant, ceux du numéro 4 ont 3 descendants et ceux du numéro 5 ont 19 descendants. Dans ce cas

$$b(A) = 1,\quad d(A) = 5,\quad q(A) = 19,\quad r(A) = \frac{\log\ 5}{\log\ 19},\quad e(A) = 5^{r(A)}.$$

G sélectionne le mode $T(r(A),q(A))$.

Exemple 3

Même départ que dans l'exemple 2. Mais les éléments du numéro 1 ont 3 descendants, ceux des numéros 2, 3 et 4 ont un descendant et ceux du dernier en ont 19. Dans ce cas

$$b(A) = 3,\quad d(A) = 5,\quad q(A) = 19,\quad r(A) = \frac{\log\ 5}{\log\ 19}$$

X sélectionne le mode $T(1,3)$, G sélectionne le mode $T(r(A),q(A))$, et $e(A) = 5^{r(A)}$.

Bibliographie

[B.P.] Benjamini,I. and Peres,Y. Tree Indexed Random Walk on Group and first Passage percolation. to appear in Probability Th and Rel. Fields.

[L] Lyons, R. Random Walk and Percolation on tree. The Annals of Probab. 1990. Vol. 18, n.3, 931–958.

Albert Benassi
Laboratoire de Mathématiques Appliquées
Université Blaise Pascal et URA 1501, CNRS
Les Cézeaux 63170 Aubière. France

Progress in Probability, Vol. 40
© 1996 Birkhäuser Verlag Basel/Switzerland

Large Deviation Principles for Random Fields on a Binary Tree

NINA GANTERT

Mathematics Subject Classification: 60J65, 60F10, 60G60, 60G18

Keywords: binary tree, ergodic transformation, large deviations, self-similarity, Brownian motion

Abstract. For a product measure on a binary tree, the shifts of the tree are ergodic transformations. We compare different large deviation principles for the corresponding empirical fields.

Résumé. Pour une mesure produit sur un arbre binaire, les translations à droite et à gauche sont des transformations ergodiques. On compare des principes de grandes déviations différents pour la suite des champs empiriques.

Introduction

Our motivation to prove large deviation principles for random fields on a tree comes from self-similarity of Brownian motion. Using the Lévy-Ciesielski construction of Brownian motion, we can represent Wiener measure as a product measure on a binary tree. The shifts of the tree to the left or to the right are ergodic transformations of Brownian motion. Starting from a "typical" path, one can reconstruct Wiener measure as the limit of the empirical fields corresponding to a sequence of these transformations. We can ask about large deviations from this convergence. There are two possibilities: one can take a fixed sequence of shifts or one can shift to the right or to the left at random. The second case was treated in [2]. Ben Arous and Tamura investigated the same model independently and got results for both cases in [1]. Note that our sequence of empirical fields does not correspond, on the lattice $\mathbb{Z}^d$, to the usual empirical field, but to the empirical distribution of a sequence of configurations we get shifting along the path of a (transient) random walk on the lattice. The same arguments go through, if we allow the random field to take values in a function space. In this way, we can describe infinite-dimensional Brownian motion as a $C[0,1]$-valued random field on a binary tree. We give here only a sketch of the arguments and refer to [1] and [2] for details and for most of the proofs.

1 Brownian Motion as a Random Field on a Binary Tree

We recall the following representation of functions in $C[0,1]_0$ which is the space of all functions X in $C[0,1]$ with $X(0) = 0$. Let the *Haar functions* φ_0, $\varphi_{n,k}$, $k = 1, 2, \ldots, 2^{n-1}$, $n = 1, 2, \ldots$ be defined as $\varphi_0(t) = 1$ $(0 \leq t \leq 1)$ and

$$\varphi_{n,k}(t) = \begin{cases} 2^{(n-1)/2} & (k-1)/(2^{n-1}) \leq t < (k-1/2)/(2^{n-1}) \\ -2^{(n-1)/2} & (k-1/2)/(2^{n-1}) \leq t < k/(2^{n-1}) \\ 0 & \text{else.} \end{cases}$$

Then $\{\varphi_0, \varphi_{n,k}, k = 1, 2, \ldots, 2^{n-1}, n = 1, 2, \ldots\}$ forms a complete, orthonormal system in $L^2[0,1]$. The *Schauder functions* e_0, $e_{n,k}$, $k = 1, 2, \ldots, 2^{n-1}$, $n = 1, 2, \ldots$, are defined as

$$e_0(t) = t, \ e_{n,k}(t) = \int_0^t \varphi_{n,k}(s)ds, \ 0 \le t \le 1, \ k = 1, 2, \ldots, 2^{n-1}, \ n = 1, 2, \ldots \ .$$

For $X \in C[0,1]_0$, set

$$h_{n,k}(X) := X\left((k-1/2)/2^{n-1}\right) - \frac{1}{2}\left(X\left(k/2^{n-1}\right) + X\left((k-1)/2^{n-1}\right)\right)$$

and

$$Y_0(X) = X(1), Y_{n,k}(X) := 2^{(n+1)/2} \cdot h_{n,k}(X), k = 1, 2, \ldots, 2^{n-1}, \ n = 1, 2, \ldots \tag{1.1}$$

Let $X^N(t) = Y_0(X) \cdot t + \sum_{n=1}^{N} \sum_{k=1}^{2^{n-1}} Y_{n,k}(X)e_{n,k}(t)$. We then have

Lemma 1.1 X^N *is the linear interpolation of X on the N-th dyadic partition of* $[0,1]$. *This implies*

$$\lim_{N \to \infty} \sup_{t \in [0,1]} |X^N(t) - X(t)| = 0.$$

Probability did not enter until now. Let now P be Wiener measure on $C[0,1]_0$.

Theorem 1. $Y_0, Y_{n,k}, k = 1, 2, \ldots, 2^{n-1}, n = 1, 2, \ldots$ *are i.i.d. random variables with distribution $N(0,1)$ under P.*

This construction of Brownian motion goes back to Lévy and Ciesielski. It is shown in [3] that one can replace the Haar functions with an arbitrary complete orthonormal system in $L^2[0,1]$.

Let $I_0 := \{0, (n,k), k = 1, 2, \ldots, 2^{n-1}, n = 1, 2, \ldots\}$. We interpret I_0 as a binary tree with root 0. The successors of 0 are $(1,1)$ and $(1,2)$, the successors of $(1,1)$ are $(2,1)$ and $(2,2)$, the successors of $(1,2)$ are $(2,3)$ and $(2,4)$ and so on.

Each function in $C[0,1]_0$ is represented by a set of coefficients $Y_0, Y_{n,k}, k = 1, 2, \ldots, 2^{n-1}, n = 1, 2, \ldots$ according to the mapping from $C[0,1]_0$ to $\mathbb{R}^{I_0}$ defined in (1.1).

The converse, however, is not true: not each element of $\mathbb{R}^{I_0}$ is a function in $C[0,1]_0$, hence not each probability distribution on $\mathbb{R}^{I_0}$ is a probability distribution on $C[0,1]_0$.

Lemma 1.3 *Let Q be a probability distribution on $\mathbb{R}^{I_0}$, $M_n := \max_{k=1,2,\ldots,2^{n-1}} |Y_{n,k}|$ and X^N as in Lemma 1.1. If*

$$\sum_{n=1}^{\infty} Q[M_n > 2^{\alpha n}] < \infty \quad \text{for an} \quad \alpha < 1/2 \tag{1.2}$$

then $Q[X^m$ *converges uniformly* $] = 1,$ *i.e.* Q *is a probability distribution on* $C[0,1]_0$.

See [2] for a proof.

Remark 1.4 We have

(i) Let $\langle X \rangle_1^{2^N} := \sum_{k=1}^{2^N} \left(X\left(k/2^N\right) - X\left((k-1)/2^N\right) \right)^2$ be the quadratic variation of X on the N-th dyadic partition. Then

$$\langle X \rangle_1^{2^N} = \frac{1}{2^N} \left(Y_0^2 + \sum_{n=1}^{N} \sum_{k=1}^{2^{n-1}} (Y_{n,k})^2 \right).$$

(ii) X is absolutely continuous with $X' \in L^2[0,1]$, i.e. X is in the Cameron-Martin space H, if and only if $Y_0^2 + \sum_{n=1}^{\infty} \sum_{k=1}^{2^{n-1}} (Y_{n,k})^2 < \infty$. In this case, we have

$$\|X\|_H^2 = \|X'\|_{L^2[0,1]}^2 = Y_0^2 + \sum_{n=1}^{\infty} \sum_{k=1}^{2^{n-1}} (Y_{n,k})^2 \ .$$

The law of large numbers implies, together with (i), that $\langle X \rangle_1^{2^N}$ converges P-a.s. and in $L^2(P)$ to 1 if $N \to \infty$. We see from (ii) that for $g \in C[0,1]$, $T_g : X \to X + g$ the translation with g and $Q = P \circ T_g^{-1}$, we have $Q \ll P$ if and only if $g \in H$.

2 Shifts on the Tree as Ergodic Transformations of Brownian Motion

For simplicity, we look here at the distribution of the Brownian bridge, which we denote again by P. Then $Y_0 = 0$ P-a.s. and $Y_{n,k}$, $k = 1, 2, \ldots, 2^{n-1}$, $n = 1, 2, \ldots$ are i.i.d. random variables under P with distribution $N(0,1)$. We set
$I := \left\{ (n,k) \mid k = 1, 2, \ldots, 2^{n-1}, n = 1, 2, \ldots \right\}$,
$\Omega = C[0,1]_{0,0} := \left\{ X \in C[0,1] \mid X(1) = X(0) = 0 \right\}$, and $X_t(\omega) := \omega(t)$. Ω is then a subset of $\mathbb{R}^I$. We will always equip $\mathbb{R}^I$ with the product topology. Denote the set of all probability distributions on Ω by $\mathcal{M}_1(\Omega)$.

We consider the mapping $T_0 : \Omega \to \Omega$, defined as

$$(T_0 X)_t = \sqrt{2}(X_{t/2} - t X_{1/2}) \qquad 0 \le t \le 1 \ .$$

T_0 is a rescaling of the left half of the path to a bridge on the unit interval: T_0 corresponds to a shift to the left of the tree, i.e. $Y_{n,k}(T_0\omega) = Y_{n+1,k}(\omega)$. In the same way, we define $T_1 : \Omega \to \Omega$ as

$$(T_1 X)_t := \sqrt{2}(X_{(t+1)/2} - (1-t)X_{1/2}) \qquad 0 \le t \le 1 \ .$$

T_1 is a rescaling of the right half of the path to a bridge on the unit interval: T_1 corresponds to a shift to the right of the tree, i.e. $Y_{n,k}(T_1\omega) = Y_{n+1,2^{n-1}+k}(\omega)$. Clearly, P is invariant under T_0 and under T_1.

We now consider the bigger space $\overline{\Omega}$, defined as $\overline{\Omega} = \Omega \times \{0,1\}^{\mathbb{N}}$, $\overline{\omega} = (\omega, \theta)$ where $\theta \in \{0,1\}^{\mathbb{N}}$. Let us look at a sequence T_θ of shifts, where $\theta \in \{0,1\}^{\mathbb{N}}$ and $(T_\theta)^k \omega := T_{\theta_k} \circ \ldots \circ T_{\theta_1}\omega$, $k \geq 1$, $(T_\theta)^0 \omega = \omega$, and define the corresponding empirical distributions

$$R_{n,\theta}(\omega) = \frac{1}{n} \sum_{k=1}^{n} \delta_{(T_\theta)^{k-1}\omega}$$

Now, there are two possibilities:

(i) We can fix $\theta \in \{0,1\}^{\mathbb{N}}$, or

(ii) We can choose θ according to the product measure λ on $\{0,1\}^{\mathbb{N}}$ (or give it another distribution).

Ben Arous and Tamura show that for each $\theta \in \{0,1\}^{\mathbb{N}}$, P is ergodic with respect to T_θ. This implies

Theorem 2.1 (Ben Arous, Tamura)
For each θ, $R_{n,\theta}(\omega)$ converges weakly to P for P-a.e. ω.

In fact, for each $\theta \in \{0,1\}^{\mathbb{N}}$ there is a decomposition of the tree I in a disjoint sequence of trees $(I_\theta^{(1)}, I_\theta^{(2)}, I_\theta^{(3)}, \ldots)$ such that $T_\theta : I \to I$ acts as a shift on this sequence: $T_\theta(I_\theta^{(1)}, I_\theta^{(2)}, I_\theta^{(3)}, \ldots) = (I_\theta^{(2)}, I_\theta^{(3)}, I_\theta^{(4)}, \ldots)$.

In [2], case (ii) is treated. Let $\overline{P} := P \times \lambda$ where λ denotes product measure on $\{0,1\}^{\mathbb{N}}$ with $\lambda[\theta_i = 0] = \lambda[\theta_i = 1] = 1/2$. Let the shift $\overline{T}$ on $\overline{\Omega}$ be defined as $\overline{T} : \overline{\Omega} \to \overline{\Omega}$ $(\omega, (\theta_1, \theta_2, \ldots)) \to (T_{\theta_1}\omega, (\theta_2, \theta_3, \ldots))$. Then, $\overline{P}$ is ergodic with respect to $\overline{T}$. This implies that for λ-a.e. θ and P-a.e. ω, $R_{n,\theta}(\omega)$ converges weakly to P, which is, of course, a weaker statement than Theorem 2.1.

Remark 2.2 Theorem 2.1 says that we can reconstruct Wiener measure with an "infinitesimal piece" of a single "typical" path around any point of the unit interval. This property characterizes fractals: the information about a fractal object is contained in a arbitrary small part of the object. Of course, we are dealing with random fractals: the invariance of $\overline{P}$ under T_0 and T_1 corresponds to a "self-similarity in distribution": $(\sqrt{2}(X_{t/2} - tX_{1/2}))_{0 \leq t \leq 1}$ and $(\sqrt{2}(X_{(t+1)/2} - (1-t)X_{1/2}))_{0 \leq t \leq 1}$ have the same distribution under P as $(X_t)_{0 \leq t \leq 1}$. We get a deterministic fractal, namely, the profile of "mount Takagi" if we set all the coefficients $Y_{n,k}$, $k = 1, 2 \ldots, 2^{n-1}$, $n = 1, 2, \ldots$ to the value 1 (see [4] for pictures).

Remark 2.3 Let Q be invariant under T_0 and T_1. Then $Q \ll P$ implies that, for each θ, $R_{n,\theta}(\omega) \to P$ for Q-a.e. ω. This means that Q cannot be reconstructed in this way, but we get P from a path which is "typical" for Q. The intuition behind this is the following: the drift of Q with respect to P is lost because of the iterated rescaling; see also Remark 1.4.

3   Large Deviations

In Section 2, we made use of the product structure of P but we did not need the marginal distribution $N(0,1)$. In the following, P will denote an arbitrary product measure on the tree. We now investigate large deviations of the convergence of $R_{n,\theta}$ to P. The decomposition of the tree used to prove Theorem 2.1 in [1] shows that, for each θ, the distribution of $R_{n,\theta}$ satisfies a large deviation principle with a rate function I_θ. In fact, since P is a product measure, the trees $I_\theta^{(1)}, I_\theta^{(2)}, \dots$ are i.i.d. and this implies that the distribution of their empirical fields satisfy a level-III-large deviation principle. We refer to [1] for details. In [2], a large deviation principle for the finite-dimensional marginals of $R_{n,\theta}$ is proved, i.e. upper and lower bounds are given for λ-a.e. θ. Here we prefer to state a large deviation principle in terms of $\overline{P}$. Let us first introduce some notation. Let $F_k(\theta, J) := \bigcup_{\ell=0}^{k-1} (T_\theta)^\ell J$ be the set of coordinates, generated by J after $k-1$ shifts according to θ, $(k \geq 1)$, $F_0(J) := J$. Let $\mathcal{F}_k(\theta, J) := \sigma(\{Y_i|\ i \in F_k(\theta, J)\})$ be the corresponding σ-field.

Consider subsets A_J of $\mathcal{M}_1(S^I)$, which are characterized in the following way: let $J \subseteq I$, J finite, $B_J \subseteq \mathcal{M}_1(S^J)$ measurable and

$$A_J := \left\{Q\,\middle|\, Q|_{\mathcal{F}_0(J)} \in B_J\right\} \tag{3.1}$$

i.e. if Q is in A_J or not depends only on the finite-dimensional marginal of Q on $\mathcal{F}_0(J)$. Then, A_J is open in $\mathcal{M}_1(\Omega)$ if and only if B_J is open in $\mathcal{M}_1(S^J)$.

We call Q *stationary* or *self-similar*, if Q is invariant under T_0 and under T_1. Let $\mathcal{M}_1^s$ denote the set of all stationary probability distributions on Ω.

The *relative entropy* $H(Q|P)|_{\mathcal{F}}$ of Q with respect to P on the σ-field $\mathcal{F}$ is defined as $E_Q[\log \frac{dQ}{dP}|_{\mathcal{F}}]$, if $Q \ll P$ on $\mathcal{F}$, and $= +\infty$ else. Now we can state the following large deviation principle:

Theorem 3.1 *Let $J \subseteq I$, J finite. Then, for all A_J of the form in* (3.1),

$$A_J \quad \text{open} \quad \implies \quad \liminf_n \frac{1}{n} \log \overline{P}\,[R_{n,\theta} \in A_J] \geq -\inf_{Q \in A_J} I_J(Q)$$

$$A_J \quad \text{closed} \quad \implies \quad \limsup_n \frac{1}{n} \log \overline{P}\,[R_{n,\theta} \in A_J] \leq -\inf_{Q \in A_J} I_J(Q)$$

where $I_J : \mathcal{M}_1(\Omega) \to [0, \infty]$ is lower semicontinuous, $I_J(Q) = +\infty$ if $Q \notin \mathcal{M}_1^s$ and for $Q \in \mathcal{M}_1^s$, we have

$$I_J(Q) = \lim_n \frac{1}{n} \int H(Q|P)|_{\mathcal{F}_n(\theta, J)} \lambda(d\theta).$$

The proof is easy from the arguments in [2]. (Actually, the lower bound is a direct consequence of the lower bound in Theorem 4.1 in [2]). The main idea in the proof of the upper bound is to represent the random variables $Z_k := \{Y_i|i \in F_k(\theta, J)\}$

as a Markov chain of order m, where $m := \max\{n|\ (n,k) \in J$ for some $k\}$ is the
"height" of J.

In contrast to [1], we insist on a rate function which is infinite on the complement of $\mathcal{M}_1^s(\Omega)$. The price we have to pay is the dependence of the rate function
on J. We give an example to illustrate this dependence:

Example 3.2 Let $I = \mathbb{R}$ and $P = \prod_{i \in I} N(0,1)$ as in Section 2. Consider A_m :
$\{Q|\ E_Q[Y_{m,1}] = E_Q[Y_{m,2}] = \ldots = E_Q[Y_{m,2^{m-1}}] \geq b\}$ $(m = 1, 2, \ldots,)$ where $b > 0$.
Since P is a product measure, we have, for all θ,

$$P[R_{n,\theta} \in A_m] = P[\overline{Y}_n \geq b]^{2^{m-1}},$$

where $\overline{Y}_n$ is the arithmetic mean of $Y_{1,1}, Y_{2,1}, Y_{3,1}, \ldots, Y_{n,1}$. So we get

$$\lim_n \frac{1}{n} \log P[R_{n,\theta} \in A_m] = -2^{m-1}h(b)$$

for all θ, where $h(x) = x^2/2$ is the rate function for the large deviations of the
arithmetic mean of i.i.d. random variables with distribution $N(0,1)$. On the other
hand, $A_m \cap \mathcal{M}_1^s(\Omega) = A_1 \cap \mathcal{M}_1^s(\Omega)$ for each m.

Theorem 3.1 says that we have to minimize the rate functional I_J over the
set of probability distributions $\mathcal{M}_1^s(S^I)$. It is therefore natural to ask about the
properties of probability distributions in $\mathcal{M}_1^s(S^I)$. We omitted Y_0, but we can
extend any stationary measure on S^I to a stationary measure on S^{I_0}. Probability
distributions in $\mathcal{M}_1^s$ are typically singular with respect to P; more precisely, if
$Q \in \mathcal{M}_1^s$ and $Q \ll P$, then $Q = P$. In particular, $Q \in \mathcal{M}_1^s$ has infinite relative
entropy with respect to P, if $Q \neq P$. We can show, though, that a *specific relative
entropy* with respect to P exists for each $Q \in \mathcal{M}_1^s$. Consider the σ-fields $\mathcal{F}_{2^n} = \sigma(\{Y_0, Y_{m,k}|\ m \leq n\})$.

Lemma 3.3 *Every $Q \in \mathcal{M}_1^s$ has a specific relative entropy $h(Q|P)$ with respect to
P:*

$$h(Q|P) \ = \ \lim_n \frac{1}{2^n} H(Q|P)|_{\mathcal{F}_{2^n}}$$

$$= \ \sup_n \frac{1}{2^n} H(Q|P)|_{\mathcal{F}_{2^n}} \in [0,\infty]\ .$$

In particular: $Q \in \mathcal{M}_1^s$, $h(Q|P) = 0 \implies Q = P$.
Further, $h(\cdot\ |P)$ is affine on $\mathcal{M}_1^s(\Omega)$.

We refer to [2] for a proof.

Can we identify a self-similar probability distribution on $\mathbb{R}^{I_0}$ with a probability distribution on $C[0,1]_0$? For $Q \in \mathcal{M}_1^s(\mathbb{R}^{I_0})$, condition (1.2) in Lemma 1.3
can be replaced by (3.2) below and we have:

Lemma 3.4 *Let $Q \in \mathcal{M}_1^s = \mathcal{M}_1^s(\mathbb{R}^{I_0})$ and*

$$\sum_{n=1}^{\infty} 2^{n-1} Q\,[|Y_0| \geq 2^{\alpha n}] < \infty \quad \textit{for an}\quad \alpha < 1/2\ . \tag{3.2}$$

Then Q is a probability distribution on $C[0,1]_0$.

In this case, we can write the σ-field $\mathcal{F}_{2^n}$ in Lemma 3.3 as $\mathcal{F}_{2^n} = \sigma(\{X_{k \cdot 2^{-n}} | k = 0, 1, \ldots, 2^n\})$, and we have an interpretation of the specific relative entropy $h(Q|P)$ as a limit of entropies on the dyadic partitions of the unit interval.

For more properties of the measures in $\mathcal{M}_1^s$ and the corresponding stochastic processes on the unit interval we refer to [2].

Of course, various generalizations are possible:

If we allow the random field to take values in a function space, we can describe infinite-dimensional Brownian motion or infinite-dimensional Ornstein-Uhlenbeck process as random fields on a binary tree: for this, we refer to [2].

The arguments carry over to trees with more branches and also to doubly infinite trees.

One might also replace the product measure with a Markov field or even a stationary random field on the tree.

Acknowledgement. I'd like to thank Gérard Ben Arous for several discussions.

References

[1] *Ben Arous, G., Tamura, Y.*: In preparation

[2] *Gantert, N.*: Self-similarity of Brownian motion and a large deviation principle for random fields on a binary tree. In: Probability Theory and Related Fields 98 (1994)

[3] *Itô, K., Nisio, M.*: On the Convergence of Sums of Independent Banach Space Valued Random Variables. In: K. Itô, Selected Papers, Springer 1987

[4] *Peitgen, H.-O., Saupe, D. (Editors)* : The Science of Fractal Images. Springer, 1988

Nina Gantert
Technische Universität
Fachbereich Mathematik
Straße des 17. Juni 136
D-10623 Berlin

List of Participants

ABRAHAM Romain
UFR de Mathématiques et informatique
Université René Descartes
45, rue des Saints-Pères
75006 PARIS
FRANCE
Email: abraham@math-info.univ-paris5.fr

ADELMAN Omer
Laboratoire de Probabilités
Université de Paris VI
Tour 46-56 - 4, Place Jussieu
75252 PARIS CEDEX 05
FRANCE

ALILI Smail
Dépt de Mathématiques -
Université de Cergy-Pontoise
8, Le Campus
95011 CERGY
FRANCE
Email: alili@u-cergy.fr

AMGHIBECH Said
URA CNRS 1378
Univ. de Rouen - UFR de Math.
76821 MONT-SAINT-AIGNAN
FRANCE
Email: Amghibec@univ-rouen.fr

ASPANDIIAROV S.
Laboratoire de Probabilités
Université de Paris VI
Tour 46-56 - 4, Place Jussieu
75252 PARIS CEDEX 05
FRANCE
Email: aspanij@oscar.proba.jussieu.fr

BABILLOT Martine
Laboratoire de probabilités
Univ de Paris VI - Tour 46-56, 3e étage
4, Place Jussieu
75252 PARIS CEDEX 05
FRANCE
Email: mbab@ccr.jussieu.fr

BENASSI Albert
Laboratoire de Mathématiques Appliquées
Université Blaise Pascal et URA 1501
CNRS, les Cézeaux
63177 AUBIERE CEDEX
FRANCE
Email: benassi@ucfma.univ-bpclermont.fr

BERCU Bernard
Université de Paris Sud
Mathématiques - Bât 425
91405 ORSAY CEDEX
FRANCE
Email: bercu@stats.math.u-psud.fr

BERTRAND MATHIS Anne
Université de Poitiers
86000 POITIERS
FRANCE
Email: Bertrand@mathrs.univ-Poitiers.fr

BESBES Mourad
Université de Versailles-Saint-Quentin
Dépt Mathématiques Bâtiment Fermat
45, avenue des États-Unis
78035 VERSAILLES CEDEX
FRANCE
Email: besbes@math.uvsq.fr

BIANE Philippe
Laboratoire de probabilités
Université de Paris VI
4, Place Jussieu
75252 PARIS CEDEX 05
FRANCE
Email: pbi@ccr.jussieu.fr

BIGGINS John
Probability & Statistics section
School of mathematics & Statistics
University of Sheffield
Sheffield S3 7 RH
ANGLETERRE
Email: J. Biggins@sheffield.ac.uk

BLONDEL Vincent
Institut de Mathématique
Université de Liège
Avenue des Tilleuls, 15
LIÈGE
BELGIQUE
Email : blondel@odyssee.inria.fr

BOUGEROL Philippe
Laboratoire de probabilités
Tour 56 - 4, Place Jussieu
75252 PARIS CEDEX 05
FRANCE
Email: bougerol@ccr.jussieu.fr

BRANDIERE Odile
Université de Marne la Vallée
2, rue de la Butte Verte
93166 NOISY-LE-GRAND
FRANCE

BROISE Anne
Université de Paris Sud
Mathématiques - Bât 425
91405 ORSAY CEDEX
FRANCE
Email: broise@topo.math.u-psud.fr

BRUNAUD Marc
Université de Paris VII-Denis Diderot
U.F.R. de Mathématiques
2, Place Jussieu
75251 PARIS CEDEX 05
FRANCE
Email: brunaud@mahtp7.jussieu.fr

CERF Raphaël
Université de Paris-Sud
Mathématiques, Bât 425
91405 ORSAY
FRANCE
Email: cerf@stats.math.u-psud.fr

CHALMOND Bernard
Université de Cergy-Pontoise
8, le Campus
95011 CERGY CEDEX
FRANCE 54506
Email: Chalmond@diam1.ens-cachan.fr

CHASSAING Philippe
URA CNRS 750
Institut E. Cartan-Nancy
B.P. 239
VANDŒUVRE LES NANCY
FRANCE
Email : Chassain@iecn.u-nancy.fr

CHAUVIN Brigitte
Université de Versailles-Saint-Quentin
Bâtiment Fermat
45, Av des États-Unis
78035 VERSAILLES - CEDEX
FRANCE
Email: chauvin@math.uvsq.fr

CHOUCROUN Francis
Mathématiques Bât 425
Université de Paris -Sud
91405 ORSAY CEDEX
FRANCE
Email: choucrou@math.u-psud.fr

CHRISTOL Gilles
Université de Paris VI
Département de Mathématiques
Tour 45-46 5ième
4, Place Jussieu
75252 PARIS CEDEX 05
FRANCE
Email: christol@mathp6.jussieu.fr

COMETS Francis
Université de Paris VII
Mathématique case 7012
4, Place Jussieu
75252 PARIS CEDEX 05
FRANCE
Email:Comets@mathp7.jussieu.fr

CRANSTON Michaël
University of Rochester
Rochester N. Y. 14627
USA

DERRIDA Bernard
Laboratoire de Physique
E.N.S
45, rue d'Ulm
75005 PARIS
FRANCE
Email: derrida@amoco.saclay.cea.fr

DUFLO Marie
Université de Marne la Vallée
2, rue de la Butte Verte
93166 NOISY LE GRAND CEDEX
FRANCE
Email: duflo@math.univ-mev.fr

FONLUPT Jean
Université de Paris VI
Tour 45-46 3ème Anal. Combinat.
4, Place Jussieu
75252 PARIS CEDEX 05
FRANCE
Email: fonlupt@ccr.jussieu.fr

COHEN Serge
Université de Versailles-Saint-Quentin
Dépt Mathématiques Bâtiment Fermat
45, avenue des États-Unis
78035 VERSAILLES CEDEX
FRANCE
Email: cohen@math.uvsq.fr

COSSART Vincent
Université de Versailles-Saint-Quentin
Dépt Mathématiques Bâtiment Fermat
45, avenue des États-Unis
78035 VERSAILLES CEDEX
FRANCE
Email: cossart@math.uvsq.fr

DANTZER Jean François
Université de Versailles-Saint-Quentin
Dépt Mathématiques Bâtiment Fermat
45, avenue des États-Unis
78035 VERSAILLES CEDEX
FRANCE
Email : dantzer@vertigo.inria.fr

DEVROYE Luc
School of Computer Science
Mc Gill University
MONTREAL
CANADA H3A2A7
Email: luc@crodo.cs.mcgill.ca

FLAJOLET Philippe
Algorithms Project - INRIA
Roquencourt
78153 LE CHESNAY
FRANCE
Email: Philippe.Flajolet@inria.fr

FOUQUE Jean-Pierre
CMAP
École Polytechnique
91128 PALAISEAU
FRANCE
Email: fouque@paris.polytechnique.fr

FRICKER Christine
INRIA
Domaine de Voluceau - Roquencourt
78153 LE CHESNAY
FRANCE
Email: Christine.Fricker@inria.fr

GAUBERT Stéphane
INRIA
Roquencourt
78153 LE CHESNAY CEDEX
FRANCE
Email: Stephane.Gaubert@inria.fr

GRAHAM Carl
CMAP
École polytechnique
91128 PALAISEAU
FRANCE
Email: Carl@cmapx.polytechnique.fr

JAFFARD Stéphane
Département Mathématiques
Université de Paris XII
Créteil CMLA, ENS.Cachan
61, avenue du Président Wilson
94235 CACHAN CEDEX
Email: sj@newaphro.enpc.fr

JOSEPH Corinne
Université de Versailles-Saint-Quentin
Dépt Mathématiques Bâtiment Fermat
45, avenue des États-Unis
78035 VERSAILLES CEDEX
FRANCE
Email: joseph@math.uvsq.fr

KOUKIOU Flora
Département de physique
Université de Cergy-Pontoise
B.P. 8428
95806 CERGY-PONTOISE CEDEX
FRANCE
Email: koukiou@u-cergy.fr

GANTERT Nina
Technische Universität
Fachbereich Mathematik
Straße des 17. Juni 136
D-10623 BERLIN
ALLEMAGNE
Email: gantert@math.tu-berlin.de

GEIGER Jochen
Universität Frankfurt
Fachbereich Mathematik
Postfach 11 19 32
D-60054 FRANKFURT
DEUTSCHLAND
Email:
geiger@mi.informatik.uni-frankfurt.de

IOFFÉ Dimitry
WIAS
39, Mohrenstrasse
BERLIN 10117
ALLEMAGNE
Email: ioffe@mamth.nwu.edu

JOFFE Anatole
Département de mathématiques
Université de Montréal
Montréal H3C3J7 QUÉBEC
CANADA
Email: joff@dms.umontreal.CA

KAHANE Jean-Pierre
Université de Paris Sud
Mathématiques Bât 425
91405 ORSAY CEDEX
FRANCE
Email: kahane@anh.math.u-psud.fr

KYPRIANOU Andrew
Probability & Statistics section
School of mathematics & Statistics
University of Sheffield
Sheffield S3 7 RH
ANGLETERRE
Email: A.Kyprianou@sheffield.ac.uk

LACROIX Jean
Laboratoire de probabilités
Tour 56 - 4, Place Jussieu
75252 PARIS CEDEX 05
FRANCE
Email: lacroix@proba.jussieu.fr

LAPEYRE Bernard
ENPC - CERMICS
La Courtine
93167 NOISY LE GRAND CEDEX
FRANCE

LASSNER François
Université de Versailles-Saint-Quentin
Dépt Mathématiques Bâtiment Fermat
45, avenue des États-Unis
78035 VERSAILLES CEDEX
FRANCE
Email: lassner@math.uvsq.fr

LE GALL Jean-François
Université de Paris VI et IUF
Laboratoire de Probabilités
Tour 56 - 4, Place Jussieu
75252 PARIS CEDEX 05
Email: gall@ccr.jussieu.fr

LEDRAPPIER François
CMAP
École Polytechnique
91128 PALAISEAU
FRANCE

LIU Quensheng
Université de Rennes I - IRMAR
Campus de Beaulieu
35042 RENNES CEDEX
FRANCE
Email: liu@levy.univ-rennes1.fr

LYONS Russell
Dept of Mathematics
Indiana University
BLOOMINGTON N4 7405
U.S.A
Email: rdlyons@iu-math.math.indiana.edu

MAZLIAK Laurent
Laboratoire de Probabilités
Université de Paris VI
Tour 46-56 - 4, Place Jussieu
75252 PARIS CEDEX 05
FRANCE
Email: mazliak@ccr.jussieu.fr

MOKKADEM Abdelkader
Université de Versailles-Saint-Quentin
Bâtiment Fermat
45, Av des États-Unis
78035 VERSAILLES - CEDEX
FRANCE
Email: mokkadem@math.uvsq.fr

MOLLOY Michaël
Dept of Computer Science
Université de Toronto
Toronto, Ontario
CANADA M5S 1A4
Email: Molloy@cs.toronto.edu

NEVEU Jacques
CMAP
École Polytechnique
91128 PALAISEAU
FRANCE
Email: neveu@paris.polytechnique.fr

NICODEME Pierre
INRIA
Roquencourt - Domaine de voluceau
78153 LE CHESNAY CEDEX
FRANCE
Email: Pierre.Nicodeme@inria.fr

PEMANTLE Robin
Wisconsin Madison University
Dept of Math, Van Vlek Hall
480 Lincoln Drive
MADISON WI 53706
U.S.A
Email: pemantle@math.wisc.edu

PERES Yuval
Dept of Statistics, 367 Evans Hall
University of California
Berkeley CA 94720
U.S.A.
Email: peres@stat.berkeley.edu

PETRITIS Dimitri
Université de Rennes I - IRMAR
Campus de Beaulieu
35042 RENNES CEDEX
FRANCE

PIAU Didier
Laboratoire de Probabilités
Université de Lyon I
43, Bd du 11 Novembre 1918
69622 VILLEURBANNE CEDEX
FRANCE
Email: piau@jonas.univ-lyon1.fr

PIERRE LOTI VIAUD Daniel
Université de Paris VI
Boite 158 - 4, Place Jussieu
75252 PARIS CEDEX 05
FRANCE
Email: Pilovi@ccr.jussieu.fr

REED Bruce
Université de Paris VI
Tour 45-46 3ème Analyse Combinatoire
4, Place Jussieu
75252 PARIS CEDEX 05
FRANCE

ROBERT Philippe
INRIA
Roquencourt
78153 LE CHESNAY
FRANCE
Email: Philippe.Robert@inria.fr

ROUAULT Alain
Université de Versailles-Saint-Quentin
Dépt Mathématiques Bâtiment Fermat
45, avenue des États-Unis
78035 VERSAILLES CEDEX
FRANCE
Email: rouault@math.uvsq.fr

ROUCAIROL Catherine
Laboratoire PRISM
Université de Versailles-Saint-Quentin
45, Avenue des États-Unis
78035 VERSAILLES
FRANCE
Email: Catherine.roucairol@prism.uvsq.fr

ROUQUÈS Jean-Philippe
Université de Versailles-Saint-Quentin
Dépt Mathématiques Bâtiment Fermat
45, avenue des États-Unis
78035 VERSAILLES CEDEX
FRANCE

SAADA Ellen
CNRS - URA Université de Rouen
Fac de Sciences, Math. site colbert
76821 MONT-SAINT-AIGNAN CEDEX
FRANCE
Email: Ellen.Saada@univ-rouen.fr

SALVY Bruno
Algorithms Project - INRIA
Roquencourt
78153 LE CHESNAY
FRANCE
Email: Bruno.Salvy@inria.fr

SIMPELAERE Dominique
Laboratoire de Probabilités
Université de Paris VI
4, Place Jussieu
75252 PARIS CEDEX 05
FRANCE
Email: ds@ccr.jussieu.fr

THOMAS Robin
School of Mathematics
Georgia Institute of Technology
Atlanta, GA 30332
U.S.A.
Email: thomas@math.gatech.edu

WERNER Wendelin
CNRS - DMI
ENS
45, rue d'Ulm
75230 PARIS CEDEX 05
FRANCE
Email: wwerner@dmi-ens.fr

STEIN Edeltrand
13, Avenue du Belvédère
78100 SAINT-GERMAIN-EN-LAYE
FRANCE

VERMET Franck
Institut de recherche Mathématiques
Université de Rennes I
Campus Beaulieu
35042 RENNES CEDEX
FRANCE
Email: vermet@levy.univ-rennes1.fr

PP - Progress in Probability

Edited by
Th. M. Liggett / Ch. Newman / L. Pitt

Progress in Probability *is designed for the publication of workshops, seminars and conference proceedings on all aspects of probability theory and stochastic processes, as well as their connections with and applications to other areas such as mathematical statistics and statistical physics.*

PP 31
H. Körezlioglu/A.S. Üstünel (Eds)
Stochastic Analysis and Related Topics
ISBN 3-7643-3666-8

PP 32
D. Nualart/M. Sanz Solé (Eds)
Barcelona Seminar on Stochastic Analysis
ISBN 3-7643-2833-9

PP 33
E. Cinlar et al.
Seminar on Stochastic Processes, 1992
ISBN 3-7643-3649-8

PP 34
M. Freidlin (Ed.)
The Dynkin Festschrift
In Celebration of Eugene B. Dynkin's 70th Birthday
ISBN 3-7643-3696-X

PP 35
J. Hoffmann-Jørgensen/J.D. Kuelbs/M.B. Marcus (Eds)
Probability in Banach Spaces, 9
ISBN 3-7643-3744-3

PP 36
E. Bolthausen/M. Dozzi/F. Russo (Eds)
Seminar on Stochastic Analysis, Random Fields and Applications
Centro Stefano Franscini, Ascona, 1993
ISBN 3-7643-5241-8

PP 37
Ch. Band / S. Graf / M. Zähle
Fractal Geometry and Stochastics
ISBN 3-7643-5263-9

PP 38
H. Körezlioglu / B. Øksendal / A.S. Üstünel (Eds.)
Stochastic Analysis and Related Topics V
The Silivri Workshop, 1994
ISBN 3-7643-3887-3

PP 39
R.J. Adler / P. Müller / B.L. Rozovskii
Stochastic Modelling in Physical Oceanography
ISBN 3-7643-3798-2

PA - Probability and its Applications

Edited by
Th. M. Liggett / Ch. Newman / L. Pitt

Probability and its Applications publishes research-level monographs and advanced graduate texts dealing with all aspects of probability theory and stochastic processes, as well as their connections with and applications to other areas such as mathematical statistics and statistical physics.

R. Carmona/J. LaCroix
Spectral Theory of Random Schrödinger Operators
1990. ISBN 3-7643-3486-X

R.K. Getoor
Excessive Measures
1990. ISBN 3-7643-3492-4

K.L. Chung/R.J. Williams
Introduction to Stochastic Integration
1990. ISBN 3-7643-3386-3

G.F. Lawler
Intersections of Random Walks
1991. ISBN 3-7643-3557-2

R.M. Blumenthal
Excursions of Markov Processes
1992. ISBN 3-7643-3575-0

S. Kwapien/W. Woyczynski
Random Series and Stochastic Integrals
Single and Multiple
1992. ISBN 3-7643-3572-6

N. Madras/G. Slade
The Self-Avoiding Walk
1992. ISBN 3-7643-3589-0

R. Aebi
Schrödinger Diffusion Processes
1996. ISBN 3-7643-5386-4

H. Holden/B. Oksendal/J. Uboe/T. Zhang
Stochastic Partial Differential Equations
A Modeling, White Noise Functional Analysis Approach
1996. ISBN 3-7643-3928-4

G.F. Lawler
Intersections of Random Walks
1996. *Softcover*. ISBN 3-7643-3892-X

N. Madras/G. Slade
The Self-Avoiding Walk
1996. *Softcover*. ISBN 3-7643-3891-1

B. Fristedt/L. Gray
A Modern Approach to Probability Theory
1996. ISBN 3-7643-3807-5